Being and Meaning

Reality and Language
in Bhartṛhari and Heidegger

Being and Meaning

Reality and Language in Bhartṛhari and Heidegger

SEBASTIAN ALACKAPALLY

MOTILAL BANARSIDASS PUBLISHERS
PRIVATE LIMITED • DELHI

First Edition: Delhi, 2002

ISBN: 81-208-1803-2

Also available at:

MOTILAL BANARSIDASS

8 Mahalaxmi Chamber, 22 Bhulabhai Desai Road, Mumbai 400 026
41 U.A. Bungalow Road, Jawahar Nagar, Delhi 110 007
120 Royapettah High Road, Mylapore, Chennai 600.004
236, 9th Main III Block, Jayanagar, Bangalore 560 011
Sanas Plaza, 1302 Baji Rao Road, Pune 411 002
8 Camac Street, Kolkata 700 017
Ashok Rajpath, Patna 800 004
Chowk, Varanasi 221 001

Printed in India

BY JAINENDRA PRAKASH JAIN AT SHRI JAINENDRA PRESS,
A-45 NARAINA, PHASE-I, NEW DELHI 110 028 AND
PUBLISHED BY NARENDRA PRAKASH JAIN FOR
MOTILAL BANARSIDASS PUBLISHERS PRIVATE LIMITED,
BUNGALOW ROAD, DELHI 110 007

DEDICATED

TO

BEINGS IN SEARCH OF BEING

FOREWORD

THE SEARCH FOR THE REAL AND THE LOGOS

The search for the Real and the Logos, Being and Meaning, is a golden thread of thought running through the philosophies of both Bhartṛhari and Martin Heidegger. The Real and the Logos are one and the same for the Indian philosopher Bhartṛhari; but it is seen primarily in the realm of timelessness. The Real and the Logos are one and the same for Heidegger too; but it is seen only in the temporal realm. Crossing over from temporal to trans-temporality is a taboo for Heidegger, whereas the Indian philosophers, in their contemplative wisdom, indulge in their undoubted assertion of the Real in its absoluteness and timelessness.

Being, hidden in every entity, emerges as Logos as the thinking mind, with its contemplative look, strips the shroud of beingness with which it is covered up. Yes, being becomes Logos, says Heidegger. But for Bhartṛhari Being, which is beginningless and beyond all forms of conceptualisation, is Logos from all eternity by its own very inner nature. For Heidegger Being becomes Logos and this process goes on and on, as the efflorescence of Being into Logos quickly vanishes, when the Being, after its tiny moment of unveiling, folds itself up back into entity. Becoming Logos is part of a flashing moment, but a powerful moment in time, which sets the human as human and being as Logos. The human is now a shepherd who summons the reality of Being, eager to unveil itself from the veil of entity, like a sheep affectionately moving to listen to the voice of its shepherd. But for Bhartṛhari the Real is inseparable from the Word and he calls this Ultimate by the name *śabda-tattva*, the Real that is the word itself.

This unveiling moment is a moment of truth. Being when it becomes Logos, truth emerges which is nothing but the mere act of unveiling (*alētheia*) as the Greek word for truth testifies. As the act of unveiling is passing, thinks Heidegger, so does the human being journeying through truth get a passing glimpse of Being, revealing itself as Logos, but only in the horizon of time, leaving the humans to an unending amazement. The Indian mind thinks differently. Beyond the horizon of time, *sat* (Being) remains as its own enlightenment and the enlightened *sat* is *satya* (truth) and *vāk* (word) at the same time. *Sat* illumined is *satya* and *satya* with its inborn dynamism of self-communication is the *vāk* (Logos) or *śabda-tattvam,* the absolutely Real ready to be *communicated. Śruti*, the listening to this unuttered Word, begins as the human search transcends the limits of temporality and listens to a voiceless sound, the unuttered Word, where sound and silence merge and give birth to Logos, the inner foundation of all languages in the world.

The contemplating human mind, listening in time, heard this Word which awakened the humans to poetry and philosophy at the same time. Fr. Sebastian Alackapally makes a journey through the rich heritages of these two philosophers hailing from East and West, from ancient and modern times respectively. Yet, he found them meeting to departing and depart to meet again. This is because the true nature of Being or the Real will be revealed to anyone, who with devotion and determination search for it, whether they see it in time or beyond time. In this encounter they see mystery beyond all comprehension, and listen to the 'unsaid' in all words which are spoken to them. Fr. Alackapally, in his enterprising and challenging study of these two authors, unveils the profoundness of their experiences representing two different times and traditions, yet fascinatingly coming closer and marching apart. I wish the author every success and invite him to search into deeper dimensions of mystery which will give a new fillip to his thinking and poetising all through his life.

Dr. V.F. Vineeth CMI
Vidyavanam Ashram
Bangalore 560 083
INDIA

ACKNOWLEDGMENTS

With a heart full of gratitude I salute the Being in whose light beings constantly unfold their way. With great affection I remember my parents and members of the family, in whose presence I began to walk before the Presence. Expressing sincere and profound gratitude I pay my respect and appreciation to all my teachers and well wishers. A special word of thanks to Frs. Lucas Vithuvattical, Thomas Aykara, Justin Koyipuram, and all my confrères, for their enthusiastic support.

Let me extend my deep felt thanks to Prof. Nico Sprokel for his guidance and personal interest in the preparation of this book. His encouraging words and compassionate dealings, and all the more his scholarship, have made my work much lighter. Also, I was fortunate to have Prof. Arij A. Roest Crollius whose academic excellence, especially in the field of Indian thought was of great help for me.

I remember with admiration Profs. Vineeth Vadakkethala, who has blessed this work with an adorning foreword, Augustine Thottakara, from whose Sanskrit knowledge I benefitted immensely, Thomas Manickam, who initiated me to the thinking of Bhartṛhari and Paulachan Kochappilly for the timely and inspiring assistance in the making of this work.

With great delight may I place this endeavour before the Presence, in the continued unveiling of Being.

Sebastian Alackapally CMI

ACKNOWLEDGEMENTS

ABBREVIATIONS

Br. Up.	*Bṛhadāraṇyaka Upaniṣad*
BT	*Being and Time*
BW	*Basic Writings*
Ch. Up.	*Chāndogya Upaniṣad*
DT	*Discourse on Thinking*
GA	*Gesamtausgabe*
HD	*Erläuterungen zu Hölderlins Dichtung*
HEP	*Hölderlin and the Essence of Poetry*
ID	*Identity and Difference*
IM	*Introduction to Metaphysics*
LH	*Letter on Humanism*
Ma. Up.	*Māṇḍūkya Upaniṣad*
PLT	*Poetry, Language and Thought*
QB	*Question of Being*
SG	*Der Satz vom Grund*
SZ	*Sein und Zeit*
Tait. Up.	*Taittirīya Upaniṣad*
TB	*On Time and Being*
VA	*Vorträge und Aufsätze*
VP	*Vākyapadīya*
VP. I	*Vākyapadīya* chapter. I
VP. II	*Vākyapadīya* chapter. II
VP. III	*Vākyapadīya* chapter. III, Part. I
VP. IIIA	*Vākyapadīya* chapter. III, Part. II
VPV	*Vākyapadīya* chapter. I with commentary
WL	*On the Way to Language*
WM	*What is Metaphysics*
WP	*What is Philosophy*
WT	*What is Called Thinking*

CONTENTS

CHAPTER ONE

ŚABDATATTVA: THE ULTIMATE REALITY

CHAPTER TWO

ŚABDATATTVA: THE *SPHOṬA* OF LANGUAGE

CHAPTER THREE

HEIDEGGER'S CONCEPT OF REALITY

CHAPTER FOUR

LANGUAGE: THE SAYING OF BEING

GENERAL CONCLUSION

BEING AND LANGUAGE IN BHARTṚHARI AND HEIDEGGER: A SYNTHESIS

GENERAL INTRODUCTION

PHILOSOPHICAL BACKGROUND OF BHARTṚHARI AND HEIDEGGER

This is a study of the concepts of Being (Reality) and Language (Meaning) as has been envisioned by Bhartṛhari, the Indian linguistic Philosopher and Martin Heidegger, a great independent German thinker. For both, the question of Being is essentially interwoven into the experience of the question of language. For them, there is no philosophy of Being without a philosophy of Language. Hence a thinking of Being is simultaneously a thinking of Language; to experience the truth of the one is necessarily to experience the truth of the other.

In Bhartṛhari's vision the language we speak is the medium of the self-expression of the ultimate Reality communicated through all meaning-bearing words. It leads us across the external appearance to the core of reality which is the source and the underlying unity beneath everything. This approach depends for its validity upon the pre-supposition that the real is a luminous Truth which needs to be dis-covered by every speaker and in every speech. The real breaks-forth (*sphuṭ*) through the medium of speech (*śabda*). This *śabda* is not merely a means to a truth or reality but it is the Truth and Reality. The awareness of this fact leads one to the realization of the meaningfullnes of Being.

In Heidegger's thinking about language the focus moves from the human activity to a source beyond man in Being or whatever expression may be used for the comprehensive reality within which humanity has its place. In his *Introduction to Metaphysics,* Heidegger sees language as not a human invention but as a pervading presence of Being on the finite human being. The dependence of human speech on the gift of Being is stressed in the *Letter on Humanism*. Heidegger gives language an essential place in the structure of Being as such. It is the original utterance (*Sage*) which Being speaks to man. Does this mean that Being itself comes to speech in human language? Obviously there is much in Heidegger that hinges on this deep connection that language is alleged to provide between Being and being.

It is language that brings Being into expression for both Bhartṛhari and Heidegger. Despite centuries of philosophy separating them, Bhartṛhari and Heidegger do seem to be committed to a unifying vision of Reality which calls us to a transformed understanding of Being and being. We intend to delve into the philosophical implications of such a vision which remains still an unexplored field of research which will lead us in good stead in continuing the philosophical inquiry into Being and its expression.

We confine ourselves to the basic theme of Being and Language throughout this study as it is envisaged by Bhartṛhari and Heidegger. In this endeavour we depend on the original works of the authors, particularly on those works which pertain to the concept of Being and Language. Emphasis is laid more on *Vākyapadiya* of Bhartṛhari, *Unterwegs zur Sprache* and other later writings of Heidegger.

Our analysis of the fundamental theme of Being and language is being explored in four chapters with an introduction and a conclusion. The introduction is an attempt to indicate the philosophical background to both authors and the development of the theme of our study. The first chapter presents Bhartṛhari's understanding of Being as *Śabdatattva*. Second chapter is an analysis of Bhartṛhari's concept of language with special emphasis on his linguistic theory of *sphoṭa*. In the same way, the third chapter deals with Heidegger's question of Being emphasizing his movement from the being to the thinking of Being, while the fourth chapter brings to light Heidegger's thinking on language. The conclusion is aimed at a synthesis of the thinking of both authors in connection with their concepts of Being and language.

The survey of the philosophical background of our authors has two parts. While the first part deals with Bhartṛhari, the second is a discussion on Heidegger.

BHARTṚHARI: THE GRAMMARIAN PHILOSOPHER

1. LIFE OF BHARTṚHARI

History is the one weak point of Sanskrit literature, being practically non-existent. Not a single systematic chronological record has survived and so complete is the lack of any data to guide us in this matter that the dates of even the most famous authors are still subject to controversy. Bhartṛhari is no exception to it. In Bhartṛhari's case much material is available, but only in the form of tales. Bhartṛhari, the grammarian was famous when I-tsing the Chinese pilgrim visited India (700 A.D.). He recorded the death of a grammarian called Bhartṛhari in 650 A.D. For long, many scholars were content to rely on I -tsings statements. But some scholars denied this view. Prof. Iyer admits Frauwallner's proposal regarding Bhartṛhari's date as more reasonable. Frauwallner has proposed the following chronological relation: Vasurāta (the *guru* of Bhartṛhari) 430-490 A.D. Bhartṛhari 450-510 A.D. and Diṇnāga 480-540 A.D.[1] Quotations from Bhartṛhari's work in the Tibetan translations of the Buddhist texts

---

[1]K.A.S. IYER, *Bhartṛhari: A Study of the Vākyapadīya in the Light of Ancient Commentaries* (Poona: Deccan college 1969), 2. Hereafter K.A.S. IYER, *Bhartṛhari.*

provide us with more reliable information regarding Bhartṛhari's date. These show that Bhartṛhari must have either lived before or been a contemporary of Diṇnāga, the Buddhist philosopher, whose dates are usually taken as c. 470-550 A.D. These and other considerations lead to the current consensus among scholars that Bhartṛhari should be dated c. 450-500 A.D.[2]

Nothing is definitely known about Bhartṛhari's life. Many stories connected with Bhartṛhari, make him a king, a brother of king Vikramaditya. One of the stories has been dramatized by Harihara in his *Bhartṛharinirveda*. In it Bhartṛhari is presented as a disciple of Gorākṣanātha from whom he learns *yoga* and renounces world. We have also I-tsing's evidence to hold the view that Bhartṛhari wavered between the ascetic life and secular life several times. I-tsing understands Bhartṛhari to be a Buddhist.[3] But this is being severely criticized by many scholars. Even a cursory glance at his famous work *Vākyapadīya* will show that he was a staunch follower of the *Vedas*. According to Ayenger, Bhartṛhari was not a Buddhist, whereas he was a strong and devout Saivite with the convictions of a Vedāntin.[4] All the popular traditions and stories regarding the grammarian Bhartṛhari seem to have no historical value whatsoever.

Bhartṛhari belongs to the tradition of the great grammarians like Pāṇini and Patañjali. Bhartṛhari for the first time undertook a thorough study of the Word-Absolute (*Śabdatattva*) and demonstrated its logical implications. And we get the absolutistic trends of the philosophy of language from his masterpiece the *Vākyapadīya*, which ranks as the principal authoritative work

---

[2]H.G. COWARD, *Bhartṛhari* (Boston: Twayne Publishers 1976), 11. Hereafter H.G. COWARD, *Bhartṛhari.*

[3]BHARTṚHARI, *Vākyapadīya* vols. I & II, tr., K.R. PILLAI (Delhi: Motilal Banarsidass 1971), xiii.

[4]V.G. AYENGAR, *Bhartṛhari's Nītiśatakam* (Thanjavur: T.M.S.M. Library 1980), xvi.

among all Sanskrit works on a philosophy of grammar.

Certain discussions on *śabda* in the *Aṣṭādhyāyī* of Pāṇini, as well as in the *Mahābhāṣya* of Patañjali and in the *Vārttika* of Katyayana also incidently introduce subjects in linguistic philosophy. These seed formulations, well before Bhartṛhari might have provided a definite background for the development of Bhartṛhari's own thinking. Hence we have a first full pledged statement and discussion of philosophy of grammar in his immortal work *Vākyapadīya*. According to R.C. Pandeya, Bhartṛhari is the first and the last original thinker of the Absolutist school of Indian philosophy of language.[5] P.T. Raju sees Bhartṛhari's thinking as a philosophy of Being developed in terms of the significance of the grammatical subject and predicate and of the meaningful word.[6] Although we find the beginnings of the concept of *Śabdatattva* (Word-Principle) in the early scriptural texts, nowhere we see a complete postulation and discussion of *Śabdatattva* as identical with *Śabdabrahman*, except in Bhartṛhari's *Vākyapadīya*. In fact, the grammarians prior to Bhartṛhari were aware of the conception of *Śabdabrahman*. Since they were content with the analysis of the language of people of their time they did not examine that transcendental and foundational principle of the human language.

2. WORKS ATTRIBUTED TO BHARTṚHARI

The following are mainly considered to be the works of Bhartṛhari, the grammarian philosopher.

1. *Mahābhāṣyaṭīkā*
2. *Vākyapadīya*, *kāṇḍas* I, II & III.

---

[5]R.C. PANDEYA, *The Problem of Meaning in Indian Philosophy* (Delhi: Motilal Banarsidass 1964), 13-14. Hereafter R.C. PANDEYA, *Problem of Meaning*.

[6]P.T. RAJU, *Structural Depths of Indian Thought* (Delhi: south Asian Publishers 1985), 17.

3. *Vṛtti on Vākyapadīya, kāṇḍa* I. & II.
4. *Śatakas* on *Nīti*, *Śṛiṇgāra* and *Vairāgya*
5. *Śabdadhātusamīkṣā.*
6. *Bhaṭṭikāvya.*

Mahābhāṣyaṭīkā is a commentary written by Bhartṛhari on the *Mahābhāṣya* of Patañjali. The manuscript which we have is only a fragment and we do not have any record of its original extent. The only available manuscript extends only to the first three *kāṇḍas* of *Mahābhāṣya* and is called *Tripāḍī. Vākyapadīya* is Bhartṛhari's great work on the Sanskrit philosophy of language. Commentary on *Vākyapadīya kāṇḍas* I and II is called *Vṛtti* or *Vivarṇa* or *Ṭīkā.* This is the oldest commentary we have on the *Vākyapadīya*. Tradition attributes it to Bhartṛhari himself. The *Śatakatṛaya*, also known as *Subhāshitatrisatī* is one of the oldest and most popular among the didactic poems in Sanskrit. According to Prof. Keith, our first great monument after Kalidasa of these kinds of verse, (Sanskrit lyric and gnomic verse) in which Indian poets admittedly excel, is to be found in the *śatakas* of Bhartṛhari[7]. *Bhaṭṭikāvya* is a work which is written mainly for the purpose of illustrating Pāṇini's grammatical rules. It is the Kashmiri writers, Somanada and Utpalacarya who give some information about Bhartṛhari's other work *Śabdadhātusamīkṣa*, a work which is extinct.

3. *VĀKYAPADĪYA*: BHARTṚHARI'S MASTER-PIECE

Vākyapadīya, the magnum opus of Bhartṛhari ranks among the principal authoritative works in Sanskrit grammar. It is the *Vākyapadīya* which is the earliest work among the grammatical treatises known to us, claiming to contain philosophical matter besides ideas relating to general linguistics and to the forms of Sanskrit language. So it is a mixture of philosophy, general

---

[7]B.A. KEITH, *A History of Sanskrit Literature* (London: Oxford University Press 1961), 175.

linguistics and grammar. *Vākyapadīya* gained vital importance among other works mainly because of its availability, completeness, new outlook and philosophical approach. Even though the text is limited to Sanskrit language and grammar only, its approach to the various aspects of language is universal in nature. For a long time there had been only a few accountable attempts towards understanding and analyzing the approach of *Vākyapadīya*. Today, however, the situation is changing and as a result, an increased interest in *Vākyapadīya* and Bhartṛhari is gaining momentum.

The *Vākyapadīya* consists of three *kāṇḍas* (sections) usually called *Brahmakāṇḍa*, *Vākyakāṇḍa* and *Padakāṇḍa*, respectively. I-tisng also approximately gives a somewhat accurate size of *kāṇḍas* I and II of *Vākyapadīya*.[8] There are scholars who treat the third *kāṇḍa* as a separate work and I-tsing's records strengthen this view. But Bhartṛhari himself in stanza 483 of *kāṇḍa* II, says that the grammatical traditions would be fully discussed in the third *kāṇḍa* making it quite clear that the third *kāṇḍa* is an integral part of the *Vākyapadīya*.[9]

The first *kāṇḍa* consists of 156 *kārikās*, (verses), the second *kāṇḍa* 485 *kārikās* and the third *kāṇḍā* 1320 *kārikās*. From the title, *Vākyapadīya* is a work dealing with sentence (*vākya*) and the word (*pada*).[10] But these two subjects are dealt with in such a comprehensive manner that ultimately the work touches upon a large number of topics. These topics are arranged in the three

---

[8]S. VARMA, *Studies in Indology* (Delhi: Bharatiya Prakashan 1981), 16. Hereafter S. VARMA, *Indology*.

[9]Because the three subsequent *kāṇḍas* (parts or sessions) of *Vākyapadīya* are seen by certain authors as three distinct books it is also called *Trikāṇḍī*, or 'Work in Three Books.' The first of the three parts is also called *Brahmakāṇḍa* or *Āgamakāṇḍa* or *Āgamasamuccaya*, the second *Vākyakāṇḍa*; the third *Prakīrṇakāṇḍa* or *Prakīrṇakakāṇḍa* or *Padakāṇḍa*.

[10]S. VARMA, *Indology*, 20.

kāṇḍas very systematically, proving the logical sequence between them. Unlike the first and second *kāṇḍas* the third *kāṇḍa* is divided into fourteen different chapters *(Samuddeśas*) named, *Bhūyodra Jāti, Dravya,Sambandha vya, Guṇa, Dik, Sādhana, Kriyā, kāla, Puruṣa, Upagraha, Saṃkhya, Liṇga,* and *Vṛtti.* A more accurate analysis of the subjects dealt within *Vākyapadīya* will include the topics of speech (*vāk*) sentence (*vākya*) and word (*pada*). Speech and sentence which form the topic of the first and second *kāṇḍas* fall directly under the purview of linguistics, while the third *kāṇḍa*, i.e., *Pada* or word, is directly related with grammar.

In fact, *Vākyapadīya*, the culminating work of the *Vyākaraṇa* (Grammar) school is a mixture of Philosophy, Linguistics and Grammar, which reveals Bhartṛhari's approach towards speech and language with regard to its basic nature. The third *kāṇḍa* is particularly concerned with the technical side of Sanskrit grammar and discusses many of the points already raised in the *Mahābhāsya* of Patañjali.

4. *VĀKYAPADĪYA*'S SUBJECT-MATTER

Verses 24-26 of the first *kāṇḍa* of *Vākyapadīya* and its *Vṛtti* give a fair description of the subject matter of the entire work. According to which the work deals with eight subjects: two kinds of meaning: that which is obtained by analysis and that which is of a fixed character; two kinds of word: that which is to be explained grammatically and that which is means of explaining it; two kinds of relation: causality and fitness to express the meaning and two kinds of result of purpose; spiritual merit and the understanding of meaning.[11]

In the first *kāṇḍa* is given the basic ideas concerning the

---

[11]BHARTṚHARI, *Vākyapadīya kāṇḍa* I, tr., K.A.S. IYER (Poona: Deccan College Postgraduate and Research Institute 1965), I.30.

concept of *Śabdabrahman*. The opening stanzas of that *kāṇḍa* endeavour to state the nature of Brahman. The treatment of *śabda* (word) or the conveyer of meaning, which is to be grammatically explained begins from verse 44 of the first *kāṇḍa*. The real word (*śabda*) is *sphoṭa*, which is to be distinguished from *nāda* or *dhvani*. However, Bhartṛhari makes use of many terms like *nāda*, *dhvani*, *prākṛtadhvani*, *vaikṛtadhvani* etc. in substitution to the term word (*śabda*). According to Bhartṛhari, a word means something in which what expresses the meaning is also included in what is expressed by it. In short the same word is regarded as the subject and the object, both at the same time. Bhartṛhari is emphatic on the point of the self expressive character of word.[12]

In the course of the exposition of the doctrine of *sphoṭa*, Bhartṛhari gives a glimpse of the different ideas regarding word which were held by different thinkers. Grammar tells us what the correct forms of words are. For Bhartṛhari, even the corrupt forms are capable of conveying meaning. The only difference between correct and corrupt forms is that the use of the former leads to spiritual merit whereas the use of the later does not.[13] Since the corrupt forms also fulfil the purpose of conveying the meaning, Bhartṛhari finds some place for their discussion in the first *kāṇḍa*.[14] Hence through the doctrines of *Śabdabrahman*, *sphoṭa* etc., the first *kāṇḍa* largely presents the outlines of the metaphysics of linguistic philosophy.

In the second *kāṇḍa* called *Vākyakāṇḍa* the discussions are largely on the nature of words and sentences. This *kāṇḍa* begins with an enumeration of looking at a sentence current in the

---

[12]G. SASTRI, *A Study in the Dialectics of Sphoṭa* (Delhi: Motilal Banarsidass 1980), 45. Hereafter G. SASTRI, *Dialectics of Sphoṭa*.

[13]VP. I.27.

[14]VP. I.141-142.

different circles of thinkers of those days.[15] It is stated that the indivisible sentence is the unit of communication and that its meaning is in the nature of intuition (*pratibhā*). The arguments of those who believe in the reality of the individual word are set forth and then refuted. Unlike the first *kāṇḍa* the whole second *kāṇḍa* is written in a polemical style. The opponent puts forward again and again his arguments in favour of the reality of the individual word; they are refuted by Bhartṛhari who holds that only the sentence and sentence meaning are real. This idea runs throughout this *kāṇḍa* even when apparently other ideas are also discussed.

The third *kāṇḍa* is the largest: in it the grammatical topics, mostly concerning words, are discussed. This *kāṇḍa* is called *Prakīrṇa* and also *padakāṇḍa*. It is the third *kāṇḍa* which deals in detail the meaning of individual words obtained by an artificial but practical analysis of the sentence.

5. *VĀKYAPADĪYA* AND COMMENTARIES

The *Vākyapadīya* is a difficult text. So commentary or interpretation of the text is essential in order to get into the depths of Bhartṛhari's thinking. The main body of the work consists of *kārikās*, of which there are a little under 2000 in all the three *kāṇḍas*. On these *kārikās* four ancient commentaries are available. The oldest of them called *Vṛtti* or *Ṭīkā* is written by Bhartṛhari himself. It covers only the first two *kāṇḍas*. Besides this, on the second *kāṇḍa* we have a commentary of Puṇyarāja. A commentary on the third *kāṇḍa* by Helārāja called *Prakīrṇakaprakaśa* is also available. Lastly we have the commentary called *Paddhati* of Vṛṣabhadeva on the first *kāṇḍa* and the *Vṛtti* thereon.

Bhartṛhari has left a vast impact over Sanskrit grammar and its

---

[15] BHARTṚHARI, *Vākyapadiya kāṇḍa* II, tr., K.A.S. IYER, (Delhi: Motilal Banarsidass 1977), 1-2.

philosophy through his crowning work *Vākyapadīya*. As I- tsing reports, Bhartṛhari's *Vākyapadīya* was one among the principal works studied by the best and most serious students, in the education curriculum of the day.[16] Even today *Vākyapadīya* is held in high esteem, in the scholarly circles.

6. *VĀKYAPADĪYA* AND THE PROBLEM OF MEANING

In Indian thought we find two main approaches to the study of the problem of meaning, namely the *khaṇḍapakṣa* and the *akhaṇḍapakṣa*. According to the *khaṇḍapakṣa* or the analytical method, a word is considered as an autonomous unit of thought and sense, and language studies are made on the basis of words, and the sentence is taken to be a concatenation of words. The idea that an individual word possesses an individual word meaning is generally current in all schools of Indian philosophy in ancient times. Thus the Nyāya school advocates meaning as a relation of words to individual objects participating in the universal which is objective. The Mīmāṃsā school replaces the Nyāya contention of an individual participating in the universal by pure universals. In their view, meaning of a word is a relation of words to universals. Whereas the Buddhists maintained the relation of words to unique individuals (*svalakṣaṇā*) which are by their nature different from others.[17]

The views of *khaṇḍapakṣa* never gave due importance to the unit of language (sentence) that it deserved. They confined their enquiry to the meaning of words. They never recognized the independence and primacy of sentences in the meaning - situation. For all of them a sentence could present nothing more than the total meaning of words constituting it. According to the *akhaṇḍ apakṣa* advocated by Bhartṛhari, the fundamental linguistic fact

---

[16] H.G.COWARD, *Bhartṛhari*, 31.

[17] R.C. PANDEYA, *Problem of Meaning*, 224.

is the sentence. He gives priority to sentences and establishes beyond doubt that words are only partial sentences; they are meaningless without reference to a sentence; words in themselves are abstractions and unreal. He defines the sentence as 'a single integral symbol' (*sphoṭa*), which is revealed by the individual letters and integral sentence (*vākyasphoṭa*) conveys meaning. And the meaning conveyed by it is an instantaneous flash or intuition (*pratibhā*).[18]

With regard to the nature of meaning, Bhartṛhari says that every word first of all means the class of that word. Thus cow means the universal of the words having the form cow; it is the universal of the 'form meant.' Later it is superimposed on the universal of the 'thing-meant.'[19] Thus the word first gives the idea of its form and then that of its referent. He says further that in all phenomenal entities there are two elements; *jāti* (universal) refers to the real element and *vyakti* (particular) to the unreal. As Helārāja points out, according to Bhartṛhari all words convey the universal. The universals are relative entities, unreal manifestations or differentiations of the highest universal; *mahāsattā*, that is Brahman. Dravid sees an advaitic overlay in Bhartṛhari's view. So it is the one and the same universal Existence which due to the diversity of particulars becomes (apparently) diversified into diverse universals. These universals are therefore ultimately non-different from existence.[20] It is to this Being, appearing in the form of different universals that all the words ultimately refer.[21]

---

[18]K. RAJA, *Indian Meaning Theories of Meaning* (Madras: Adyar Library and Research Centre 1977), 9. Hereafter K. RAJA, *Meaning Theories*.

[19]BHARTṚHARI, *Vākyapadīya*, Chapter III, Pt. I, tr., K.A.S. IYER (Poona: Deccan college 1971), III.1.6.

[20]R.R. DRAVID, *The Problem of Universals in Indian Philosophy* (Delhi: Motilal Banarsidass 1972), 263. Hereafter R.R. DRAVID, *Universals*.

[21]VP. III.1.33.

What is worthy of emphasis is that whether the universal is looked upon as the meaning of all words or whether substance is so looked upon, it is ultimately Brahman (*Śabdatattva*) which turns out to be the meaning (*artha*) of all words. Thus the discussion on the problem of meaning by Bhartṛhari shows his extra ordinary linguistic and philosophical acumen. With these brief notes on Bhartṛhari we shall now move to Heidegger.

HEIDEGGER:
THE PHILOSOPHER OF BEING AND LANGUAGE

1. LIFE AND PHILOSOPHICAL BACKGROUND

Martin Heidegger is widely recognized as one of the greatest philosophers of the twentieth century. He was an original thinker, a critic of technological society and a leading ontologian of his time. Born at Messkirch in the Black Forest in 1889 and, as a Roman Catholic, well acquainted with Thomistic thought from his early youth onwards, Heidegger received his first philosophic training in the Neo-Kantian school. At the University of Freiburg he studied Catholic theology and medieval Christian philosophy. His interest in philosophy arose as a result of an intensive study of the late 19th century Catholic philosopher Franz Brentano's *Von der mannigfachen Bedeutung des Seienden nach Aristoteles* (1862; *On the Manifold Meaning of Being According to Aristotle*). From the time of his early study of Brentano also stems his enthusiasm for the Greeks, especially the pre-Socratics. Hence the philosophy of Heidegger is very much dependent upon the philosophers prior to Socrates, upon Plato and Aristotle. He was particularly greatly influenced also by several 19th and early 20th century philosophers: by Søren Kierkegaard and Friedrich Nietzsche, founders of Existentialism; by Wilhelm Dilthey, and by the founder of phenomenology, Edmund Husserl.

Heidegger started teaching at the University of Freiburg during the winter semester of 1915 and earned his habilitation through a study of Thomas of Erfurt, whom he thought to be the 13th century British Franciscan philosopher Duns Scotus. In this position, as a colleague of Husserl, Heidegger was expected to carry the Phenomenological movement further along in the spirit of Husserl who was also his master. But he broke away to fashion his own philosophy. In 1927 he astonished the German philosophical world with his magnum opus *Sein und Zeit*[22] (*Being and Time*) - a work which in spite of its intriguingly difficult style was immediately felt to be of prime importance. It strongly influenced Jean-Paul Sartre in France and other Existentialists, and, despite Heidegger's protestations, he was classed, on the strength of this book, as the leading atheistic Existentialist.

Heidegger's thought has had strong influence on trends in philosophy ranging from existentialism through hermeneutics to deconstruction, as well as on the fields of literary theory and theology. His writings have had an immense impact not only in Europe and the English-speaking world, but in Asia as well.[23] Heidegger transformed mainstream philosophy by defining its central task as asking the "question of Being." His lofty ambition was to rejuvenate philosophy by clearing away the conceptual rubbish that has collected over Western history in order to recover a clearer, richer understanding of what things are all about. Since this calls for appropriating the underlying ideas that have formed the Western culture, his thought weaves together many different historical strands.

2. HEIDEGGER'S METHODOLOGY

In close contact with Husserl, the founder of phenomenology, Heidegger developed his own method of the interpretation of the

---

[22]M. HEIDEGGER, *Sein und Zeit* (Frankfurt am Main: Vittorio Klostermann, GA. 2, 1977).

[23]See, G. PARKES, ed., *Heidegger and Asian Thought* (Delhi: Motilal Banarsidass 1987). Hereafter G. PARKES, *Asian Thought*.

texts and ideas of great philosophers of the past, and of the exposition of systematic problems to which the tradition, from the Greeks to Husserl, and other eminent thinkers of the present age, gave rise. According to Heidegger, the concept of "Being" is the most universal one, as was also realized by Aristotle, Thomas and Hegel; and its universality goes beyond that of any "genus." At the same time it is obscure and indefinable; "Being" cannot be comprehended as anything that is (*Seiendes*); it cannot be deduced from any higher concepts and it cannot be represented by any lower ones; "Being" is not something like a being. Yet "Being" seems somehow an evident concept. We make use of it in all knowledge, in all our statements, in all our behaviour towards anything that "is," in our attitude towards ourselves. We are used to living in an "understanding of Being" (*Seinsverständnis*), but hand in hand with it goes the incomprehensibility of what is meant by "Being." To-day, and in fact throughout the last century, the problem of "Being" has fallen into oblivion.

According to Heidegger, it is the phenomenological and not the scientific method that unveils man's ways of Being. Thus, in pursuing this method, Heidegger comes into conflict with the dichotomy of the subject-object relation, which has traditionally implied that man, as knower, is something (some-thing) within an environment that is against him. The deepest knowing, on the contrary, is a matter of *phainesthai* ("to show itself" or "to be in the light"), the word from which phenomenology, as a method, is derived. Something is just "there" in the light. Thus, the distinction between subject and object is not immediate but comes only later through conceptualization, as in the sciences.

3. HEIDEGGER'S PROJECT

Heidegger's *Being and Time* made at once a profound impression upon the philosophically-minded public, even outside the sphere of the trained philosophers, and was soon considered to be a landmark in philosophic studies. The aim of this great work, and indeed of all of Heidegger's publications, is the re-awakening of the question, What is meant by "Being"? This problem belongs to the tradition of European philosophy from the Greek Philosophers Anaximander and Parmenides onwards; more than that, it was its central problem. In Heidegger's view, it guided the exertions of the greatest among the pre-Socratic thinkers as well as those of Plato and of Aristotle. But after Aristotle it ceased to be the thematic problem of a genuine philosophic enquiry. The achievement of elucidation attained until Aristotle affected vitally the medieval discussion of the problem and the whole of the Christian theological outlook; and through many changes the tradition of the problem kept alive down to Hegel's *Logic*.

Heidegger's aim in *Being and Time* is to revive the question about the meaning of Being, in the sense in which it was the guiding problem of Greek thought until Aristotle and its more or less expressed theme of enquiry. In this respect he takes the Greek thinkers as his model. But he deviates from them fundamentally in his starting-point. They reflected upon the things encountered in the world, that could be seen and thus known. And the thing that was perceived and about which statements could be made in various relevant respects, i.e., by way of "categories," was their paradigma. Heidegger's starting point is not the perceptible things, but what he terms: Dasein (There-being),[24] a phenomenon fundamentally, i.e., in its ontological structure, not analysed by the Greeks or ever since in later philosophic tradition.

Heidegger's endeavour is to give an analysis of the "*existentialia*"and of the "existentialistic" structure of Dasein in a way which the Greek thinkers developed the "categories" of a thing that "is." But this analysis, profound and original as it is, is to him nothing but the starting-point. It is from this new angle that

---

[24]Because the word Dasein is already familiar to the English readers of Heidegger, we prefer to leave it unitalized and often untranslated.

he intends to unfold the problem of "Being"afresh. The Being question lies behind the obviousness of everyday life and, therefore, also behind the empirical questions of natural science. It is usually overlooked because it is too near to be grasped in everyday life. One might say that the whole prophetic mission of Heidegger amounts to making each man ask that question with maximum involvement. Whether he will ever arrive at any definite answer or not is of secondary importance. In Heidegger's later writings, he suggests that there are possibilities of redemption by "thinking of Being" and, thus, again coming closer to Being.

4. APPROACH TO BEING THROUGH DASEIN

In order to clarify the conditions that make possible an understanding of Being, then, *Being and Time* begins with an analytic of Dasein. But Heidegger notes that traditional interpretations of human existence have been one-sided to the extent that they concentrate on our ways of existing when we are engaged in theorizing and detached reflection. It is this narrow focus on the spectator attitude that leads to the picture, found in Descartes, of the self as a mind or subject representing material objects - the so-called subject-object model. In order to bypass this traditional picture, Heidegger sets out to describe Dasein's "average everydayness," i.e., our ordinary, pre-reflective agency when we are caught up in the midst of practical affairs. The phenomenology of everydayness is supposed to capture the totality of human existence, including our moods, our capacity for authentic individuality, and our full range of involvements with the world and with others.

The analytic of Dasein is also an ontological hermeneutics to the extent that it provides an account of how understanding in general is possible. The result of the analytic is a portrayal of human existence that is in accord with what Heidegger regards as the earliest Greek experience of Being as an emerging-into-presence (*physis*): to be human is to be a temporal event of self-manifestation that lets other sorts of entities first come to "emerge and abide" in the world.[25] From the standpoint of this description the traditional concept of substance simply has no role to play in grasping humans. Heidegger holds that there is no pre-given human essence. Instead humans as self-interpreting beings just are what they make of themselves in the course of their active lives. Thus, as every day agency Dasein is not an object with properties but rather is the "happenings" of a life course "stretched out between birth and death." Dasein finds itself thrown into a world not of its choosing, already delivered over to the task of living out its life in a concrete context. Again Dasein is always already taking some stand on its life by acting in the world.

Heidegger realized that Dasein differed ontologically from all the things which are not Dasein in essential respects. These things, when they are there by nature, are termed *vorhanden* ("existent" in the usual sense of the word, literally: before one's hand, at hand, present); and when they are made by men, such as utensils, they are termed *zuhanden* close at hand, at one's disposal. Dasein is always my own Dasein. It cannot be ontologically grasped as the case or the example of a genus of beings, as can be done with things that are *vorhanden.* This by itself causes considerable difficulties for the adequate ontological exposition. Besides, the being of the kind of Dasein is in its Being concerned about its Being and behaves towards its Being as towards its own possibility. It chooses and decides and it may gain or may lose itself, in as far as its Being is concerned. All this cannot be said of things that are *vorhanden.*

Two fundamental modes of Being, authenticity and unauthenticity, are distinguished, both of them depending on the fact that Dasein is essentially always my own. Of all the things that are *vorhanden* it can be stated that they are of a special 'genus,' e.g. a house or a tree, and that they have special

---

[25]See, R. AUDI, ed., *The Cambridge Dictionary of Philosophy* (Cambridge: University Press 1995), 317-318.

"qualities." In other words, their "essence" is always ascertainable. In contrast to them, the characteristics of Dasein are not "qualities," but possible ways of "Being." Therefore the term Da-sein is to express not its "essence," but "Being"; it means "There Being." To distinguish further the kind of Being, peculiar to Dasein, from all *Vorhandenheit*, the term "existence" is applied exclusively to it. And the fundamental characteristics of Dasein, corresponding to but in opposition to the categories of *Vorhandenheit,* therefore termed *existentialia.*[26]

5. QUESTION OF TEMPORALITY

If the aim of BT is the reawakening of the question: what is meant by Being? and if its starting-point is an ontological analysis of Dasein, the one main problem not yet considered is that of Time. The problem of Time is the link between the analysis of Dasein and the revival of the question of the meaning of Being. Here again Heidegger's approach seems to be in vital contrast to that of the Greeks and the ontological tradition which they initiated. In Heidegger's view, the meaning of "Being" is intimately bound up with the phenomenon of Time and has been bound up in this way since the beginning of philosophic thought. For the Greeks the definition of the Being of the things that are were envisaged in their Being as "present." This basis of the interpretation of the things in their Being has never been fundamentally questioned.

The temporality of Dasein, with its relations to future, past and present - to what Heidegger terms the three "ecstasies" of temporality - opens up the "horizon" for the question about "Being" in an entirely new way so that this question can be re-asked only after this analysis of temporality. The relevance of Existence be-comes clearer here through its prevalent relationship to the future; and it may be said that all the dominant

[26]M. HEIDEGGER, *Being and Time*, tr., J. MACQUARRIE & E. ROBINSON (NewYork: Harper & Row 1962), 67-68.

characteristics of Dasein gain their fuller significance in the interpretation of temporality. The aim of the exposition of the temporality of Dasein is to gain an insight into the nature of Time itself, an insight which, in Heidegger's view has hitherto not advanced substantially beyond Aristotle's interpretation of Time in the *Physics*. The explication of Time as the "transcendental horizon" for the problem of "Being" was lead to the aim: the analysis of what is meant by Being.

In the decades following the publication of BT, Heidegger's *Seinsfrage* shifted its focus. The epoch of BT and of what is often referred to as the earlier Heidegger is followed by the epoch of later Heidegger. The emphasis on the primacy of Being became more and more pronounced with the unfolding of Heidegger's way of thinking, especially in the later works. An analysis of Heidegger's shift of emphasis from being to Being forms the subject matter of our third chapter.

6. HEIDEGGER'S LANGUAGE

Heidegger's later writings attempt to fully escape the subjectivism he sees dominating Western thought from its inception up to Nietzsche. In his writings after the second war, Heidegger replaces the notions of resoluteness and political activism with a new ideal of letting-be or releasement (*Gelassenheit*), a stance characterized by meditative thinking, thankfulness for the "gift" of Being, and openness to the silent "call" of language. Language comes to be understood as the medium through which anything, including the human, first becomes accessible and intelligible. Because language is the source of all intelligibility, Heidegger says that humans do not speak, but rather language speaks us - an idea that became central to post-structuralist theories.

Heidegger's understanding of language in BT arises out of his investigation of Dasein, the being whose mode of Being is to inquire concerning its own Being. The third of the *existentialia* of

Dasein co-original with state-of-mind (*gestimmte Befindlichkeit*) and understanding (*Verstehen*), is speech (*Rede*).[27] Speaking is the signifying articulation (*Artikulation, Gliederung*) of the Being-in-the-world in the way in which it is understood. Dasein is discourse (*Rede*) in the sense that we are always articulating - or "addressing and discussing" - the entities that show up in our concernful absorption in current situations. The understanding, arising for human Dasein in the 'There' in which it is 'placed,' expresses itself as speech; and the totality of the significances articulated by understanding and interpretation comes to 'word.'

Here again one notices a profound change in Heidegger's earlier understanding of language and his later understanding of language which is presented mainly in his essays collected under the title *Unterwegs zur Sprache*.[28] Hence over the course of the two major periods of his career Heidegger has developed two distinct but related understandings of the nature of language. In each period the problem of the nature of language is intimately connected with his principal problematic: the problem of the meaning of Being. In the first period represented by BT, Heidegger approaches the problem of the meaning of Being by means of a phenomenological investigation of the existence of the individual human being. This being which Heidegger characterizes by the term Dasein is philosophically distinguished as the being which has the capacity to inquire concerning its own Being. This investigation leads him to maintain that the Being of Dasein is a structural unity constituted according to three fundamental modes of equal importance: under the mode of discourse, language is found to be constitutive of the existence of the individual human being.

After the BT period, Heidegger passes through his famous

---

[27]BT. 172.

[28]M. HEIDEGGER, *Unterwegs zur Sprache* (Frankfurt am Main: Vittorio Klostermann, GA.12, 1985).

turn, in which he puts aside the phenomenological method in favour of his wide ranging 'way of thought' (*Denkweg*). Influenced by his meditations on the sayings of the Greek pre-Socratics, Parmenides and Heraclitus, and on the works of the modern German poets, especially Hölderlin, he develops the 'way of thought' according to the single guiding theme that thought must always strive to stand in the light of the truth of Being. Thus in this period Heidegger enters into the meditation of Being as such. This problematic which leads him past the problems of the Being of man and the Being of beings , brings him to understand Being concretely as the "Event of Appropriation" (*Ereignis*).

The understanding of Being as Appropriation arises out of his meditation of the Greek pre-Socratic usage of the word *logos*. In his retrieval of this key Greek word Heidegger finds that it does not merely signify 'language' or 'reason,' but indicates the primal gathering of beings in Being. Accordingly in the later period language, as founded in *logos* transcends the individual human being, and is in fact the gathering of the manifestness of beings within the enduring presence of Being.

In the later works Heidegger has turned from the analysis of Dasein to the more general investigation of Being as such. Accordingly his understanding of the nature of language, as it appears in the later works develops out of this new perspective. His understanding of language indeed finds its roots in his thinking of the truth of Being. Here Heidegger indicates that in man's special relationship to Being he is active with regard to the revealing and concealing of beings. Man, as engaging in this particular activity, must have a particular capability - the 'word.' This intimation of the role of the 'word' as proposed by Heidegger indicates that speech is directly involved in the disclosure of the Being of beings.

Heidegger asserts that the role of the 'word' and of speech is very special, to be distinguished from the role of everyday language, which is a derivative from the speaking of the original 'word.' Heidegger's understanding of language, intimately intertwined with his understanding of Being and the being of man, thus undergoes a change as it passes through early to the later period of his philosophic thinking. Our fourth chapter attempts the development of this theme. His two apparently divergent understandings of the nature of language are reconciled on the basis of the relationship which they maintain within and to the stages of the unfolding of his philosophical problematic - the question of Being.

CHAPTER ONE

ŚABDATATTVA: THE ULTIMATE REALITY

1. INTRODUCTION

Bhartṛhari begins his treatment of language with the identification of *śabda* (word)[1] with Brahman.[2] In his vision Brahman is the essence, inner meaning or *sphoṭa* (a latent unitary medium which is the real meaning-bearing whole) of every word. He calls this inner principle of word *Śabdatattva* (Word-Principle or Word-Essence) which is the underlying unity beneath everything and is considered to be the Supreme Reality in the Philosophy of the Grammar School. In fact, the English expressions often fail to convey the full connotations of the term *Śabdatattva*. In the absence of any other suitable expression, G. Sastri suggests the term Eternal Word (*Verbum Eternum*) or Supreme Word which very well corresponds to the original conception of *logos*.[3] In the present chapter we propose to analyse the nature of *Śabdatattva* as has been envisaged by

---

[1]For an overview of the development of the notion of *śabda*, see Appendix. For a useful introduction to the place of word in the Hindu Scriptures, see also, A.A. ROEST CROLLIUS, *Thus Were They Hearing*: *The Word in the experience of Revelation in Qur'ān and Hindu Scriptures* (Roma: Università Gregoriana 1974).

[2]Throughout the text italic type of the term is avoided.

[3]G. SASTRI, *The Philosophy of Word and Meaning* (Calcutta: Sanskrit College 1959), 1. Hereafter G. SASTRI, *Philosophy of Word and Meaning*.

Bhartṛhari. Attempt is made to delve in to Bhartṛhari's way of looking at Reality along with an evaluation of the realistic and idealistic orientations of his thought. We also take up a discussion of the Reality in the context of its relation to time, language and meaning. The chapter is concluded with an appreciation of the Linguistic Absolutism of Bhartṛhari.

2. QUEST FOR THE REAL

The whole question regarding the nature of the Absolute has always remained an unresolved riddle. Philosophers in their attempt to give possible explanations to the nature of the Absolute gave birth to various schools of philosophy both in the East and in the West. Thus, the Advaitins of East and the Monists in the West held that reality is one and non-dual. If reality is one and non-dual, it definitely calls into question the multiplicity of beings which we experience everyday. By contrast, the Saṃkhyans of Indian thought and Atomists from the West maintained that Reality is pluralistic in nature. This view too is not without any difficulty. For, their conclusion fails in explaining the ultimate unity of all in one. Hence Rāmānuja's (c.1027) qualified non-dualism and the whole of Thomistic philosophy take an intermediary position, as a synthesis of the previous views. They acknowledge the one along with the co-existence of many.

In the Indian context the Sāṃkhya-Yoga, Nyāya-Vaiśeṣika and Pūrva-Mīmāṃsa systems are examples of dualistic or pluralistic Metaphysics, while the Vedānta of the school of Śaṇkara (c.725) is the most outstanding example of monism. Besides Śaṇkara's Advaita (non-dualism) which is usually known as *Brahmādvaita,* there have been several other attempts at explaining the phenomenal universe of diversity from one absolute principle without any second. Among them Bhartṛhari's doctrine of the *Śabdatattva-vāda* or *Śabdādvaita-vāda* (Linguistic Absolutism) is a novel doctrine altogether, though its origin is traceable in the Vedic *Saṃhitās* and in the conception of the sacred syllable *aum* in the *Upaniṣhads*.

3. THE REAL IS *ŚABDATATTVA*

Bhartṛhari conceives Reality as *Śabdatattva.* The compound *Śabdatattva* can be dissolved in three different ways *śabdasya tattvam* - Brahman is the essence of *śabda*;[4] *śabdāḥ tattvam asya* - whose very essence is the word;[5] and *śabda eva tattvam* - word essence.[6] Since Bhartṛhari holds that Brahman is the supreme form of the undifferentiated *vāk* (word)[7] and the underlying reality of the world of *śabda* and the multiplicity manifested by it, the derivation *śabdasya tattvam* seems appropriate. Bhartṛhari considers Brahman which is of the nature of *śabda,* in its real form is also the nature of consciousness.[8] K.A.S. Iyer, referring to *Vṛtti*,[9] explains the reason why Brahman is considered to be the Word-Principle:

> ... the universe is really Brahman who creates all objects and phenomena in the form of words. Just because all that Brahman creates has the form of word, therefore Brahman itself must be of the nature of word. The fact that Brahman is called *akṣara*, phoneme is also an indication of its being the word-principle. As everything else, the phonemes also emerge out of Brahman. They exist potentially within the individual, as one with the self, without any sequence. ...

---

[4]S. SARMA, "Some Aspects of Bhartṛhari's Philosophy," *Sri Venkateswara University Oriental Journal* 5 (1962) 39.

[5]BHARTṚHARI, *Vākyapadīya with Vṛtti,* tr., K.A.S. IYER (Poona: Deccan College 1965), I.1. Hereafter VPV.

[6]G. SASTRI, *Philosophy of Word and Meaning,* 1.

[7]*pratyastamitabhedāya yad vāco rūpam uttamaṃ /*
yad asmin eva tamasi jyotiḥ śuddham vivartate // VP.I.18.

[8]*prāptarupavibhāgāya yo vācaḥ paramo rasaḥ /*
yat tat puṇyatamaṃ jyotis tasya mārgo'yam āñjasaḥ // VP.I.12.

[9]There are four ancient commentaries available on *Vākyapadīya. Vṛtti* is oldest of these commentaries on the first and second *kāṇḍas* (section) and is traditionally considered to be written by Bhartṛhari himself. Other commentaries include: *Prakāśa* by Puṇyarāja (c.1000) on the second *kāṇḍa*, *Prakīrṇakaprakāśa* by Helārāja (c.980) on the third *kāṇḍa* and *Paddhati* of Vṛṣabhadeva (c.600) on the first *kāṇḍa.*

> they are uttered. Brahman is called *Śabdatattva* because all phenomena assume the form of the word and also because it manifests itself as the uttered phonemes for the purpose of communication. The phonemes are a kind of overflow of the subtle word within.[10]

The *Śabdatattva* of Bhartṛhari differs from the upaniṣadic conception of *Śabdabrahman.* According to the upaniṣadic vision, there are two Brahmans to be known: Word Brahman and the Supreme, it is this Supreme Brahman that man reaches when he is proficient in the word Brahman.[11] Whereas for Bhartṛhari *Śabdatattva* is the Absolute and there is no distinction between the *Śabdabrahman* and the Supreme. The speech essence (*Śabdatattva*) is the ultimate Reality which has neither beginning nor end and is unchanging. Bhartṛhari expounds this view in the very first *kārikā* (verse) of *Vākyapadīya* thus:

> *anādi nidhanaṃ brahma śabdatattvaṃ yadakṣaraṃ /*
> *vivartate arthabhāvena prakriyā jagato yataḥ //*
>
> That beginningless and endless One, the imperishable Brahman of which the essential nature is the Word, which manifests itself into objects and from which is the creation of the Universe.[12]

Bhartṛhari conceives the ultimate Reality as One. It is of the nature of word and from it are manifested all objects and the whole cosmos. So, this very first *kārikā* of *Vākyapadīya* would mean that the whole phenomenon of material existence is only an appearance (*vivarta*) of the speech principle which is identical with the ultimate Reality, Brahman. Conception of Brahman as the word principle or the identification of Brahman with the *Śabdatattva* forms the central theme of *Vākyapadīya* which gives a kind of unity to the whole text. *Vṛtti* clarifies Bhartṛhari's stand point saying that the Word-Absolute has the dual aspects of unity

---

[10]K.A.S. IYER, *Bhartṛhari*, 101.

[11]S. RADHAKRISHNAN, *The Principal Upaniṣads* (London: George Allen & Unwin 1953), 833. All the references from the *Upaniṣads* are taken from this edition.

[12]VP. I.1.

and diversity, but, of these the former alone is real, while the latter is merely apparent.[13] The second *kārikā* also develops the same vision:

> *ekam eva yadāmnātam bhinnam śāktivyapāśrayāt /*
> *apṛthktve'pi śaktibhyaḥ pṛthaktveneva vartate //*
> Which (Brahman) though described in the Vedas as one is divided on the basis of its powers, and although it is not different from its powers appears to be different.[14]

Hence the *kārikā* makes it clear that the speech essence which is the ultimate Reality has neither beginning nor end and is unchanging, but on the basis of its various powers such as time, which, though in essence identical with it, seem to be different, the phenomenal world appears as evolutionary and pluralistic. It manifests itself without loosing its one-ness. The opening verses of *Vākyapadīya*, thus present us with a crucial statement about the transcendent Reality which is beyond all limitations of time and space. It is labelled significantly as Brahman as well *Śabdatattva*. It is both *anādi* (beginnigless) and *a-nidhana* (endless). It is *akṣara* (imperishable). *Akṣara* also signifies phonemes, indicating that the very essence of Brahman itself is the Word-Principle. While it refers to the immanent nature of *Śabdatattva*, *anādinidhanam* refers to the transcendent aspect of the *Śabdabrahman.*

Time is an inherent power of the Absolute which exerts its influence in bringing about the powers of *Śabdatattva*.[15] It does not mean that the *Śabdatattva* is limited by time, but the eternal timeless appears as changing owing to the working of time factor.[16] As stated in the fourth *kārikā* Brahman is the non-dual

---

[13]VPV. I.1.

[14]VP. I.2.

[15]*adhyāhitakalāṃ yasya kālaśaktim upāśritāḥ /*
janmādayo vikārāḥ ṣaḍ bhāvabhedasya yonayaḥ // VP.I.3.

[16]K. RAJA, *Meaning Theories*, 146-147.

source of all multiplicity that comes into existence. In other words the substratum of the manifold, Brahman includes all multiplicity.[17] In this context G. Sastri suggests that the first five verses of *Vākyapadīya* should be taken as a syntactically connected whole describing the transcendent and immanent aspects of the ultimate Reality.[18]

In fact, Bhartṛhari is ever conscious, throughout the *Vākyapadīya*, of the ultimate Word-Principle, the *Śabdatattva*, Brahman out of which the whole cosmos and our experience of it consisting of an infinite variety of cognitions, objects and words expressive of them are manifested. That runs like a thread through his work and gives it a kind of unity. Whatever he speaks, he somehow or other connects with the inner *Śabdatattva*.

The *Śabdatattva* as the source of all manifestations appears in the form of the manifold subject, object and the process of experience.[19] Since production and destruction take place in time and space, negation of these connotes that Brahman transcends all change and that it is eternal and all-pervasive.[20] As it is non-spatial, it is also indivisible. It is eternal in the sense that it is the timeless reality of all that exists in time. This undifferentiated and unchanging Reality, beyond determinations of thought and words is non-relational as there is no other entity with which it could be related or associated. It is immutable as it transcends space, time and causality. It has neither a beginning nor an end. Bhartṛhari also refers to that aspect of the eternal reality which transforms itself into speech, as words and their meanings and into matter as the different subjective and objective elements of the phenomenal

---

[17] *ekasya sarvabījasya yasya ceyam anekathā /*
bhoktṛbhoktavyabhedena bhogarūpeṇa ca sthitiḥ // VP. I.4.

[18] G. SASTRI, *Dialectics of Sphoṭa*, 89.

[19] VP. I.4.

[20] BHARTṚHARI, *Vākyapadīya*, Chapter III, Part I., tr., K.A.S. IYER, (Poona: Deccan College 1971), III.2.18.

world. Nevertheless, these transformations do not affect the true essence of Reality and it remains unchanged while persisting in all its modifications.

Bhartṛhari devotes a considerable portion of the first *kāṇḍa* (section) of the *Vākyapadīya* to establish *Śabdatattva* on the basis of tradition (*āgama*), inference (*anumāna*) and perception (*pratyakṣa*). *Āgama* is the most authoritative of all means of knowledge according to Bhartṛhari. He presents many passages from ancient texts which proclaim *śabda* as the principle from which everything originates in which it exists and into which it merges on destruction. Brahman is the essence of *śabda.* It creates the universe through *śabda*. Universe and Brahman which are identical ultimately, are manifested by *śabda*. Bhartṛhari also finds support from the knowers of tradition who hold that everything is a modification of *śabda* and that the universe is the first transformed from the *śabda.*[21]

Bhartṛhari proves the nature of the *Śabdatattva* not only with the help of tradition but also reasoning (*anumāna*). He adopts the philosophic dictum used by many schools of philosophy, namely, that all modifications partake of the nature of their cause (*vikārāṇām prakṛtyanvayitvam*). It is noticed that all objects contain their material cause. Jewels shaped out of gold are golden in nature and vessels made of clay partake of the nature of clay which is their material cause. Since the material cause persists in all its effects, it is possible to infer the nature of the cause on the basis of the element found consistently in all the effects. In other words, since cause continues to exist in all its effects, that which is common to all effects may be postulated to be their cause.

Bhartṛhari argues that the cause of the universe must be of the nature of *śabda* because all objects are known through their words and because all cognitions are invariably intermixed with

---

[21]*śabdasya pariṇāmo 'yam ity āmnāyavido viduḥ /*
chandobhya eva prathmam etad viśvam vyavartata // VP. I.120.

śabda. That all objects of knowledge are of the nature of *śabda* is evident from the fact that the knowledge of the word is indispensable for knowing the object. Unless the word is known, the object cannot be comprehended. Also, when the name is known, the object is held to be known irrespective of its existence or non-existence in the external world. Since the object that shapes the cognition is of the nature of *śabda*, cognition is also of the same nature. Bhartṛhari declares that there does not arise any cognition which is not fused with *śabda*. All knowledge appears intertwined with *śabda*.[22] Hence the object is really not different from the word. It is the word which has become the object.[23]

Bhartṛhari gives an elaborate account of the all-pervasive role of *śabda* in the universe. *Śabda* is not only the creator of the universe but also the revealer of the objects contained in it. All differentiation of the world depends on *śabda* because it distinguishes and interperts that would otherwise be a mere undifferentiated mass of existence. Objects are recognized and given name and form through *śabda* which also gives form to cognitions.[24] All cognitions are intertwined with *śabda*. Even the vague perception of objects whose specific features are not noticed is known by pronouns like 'this' or 'that.' In other words, the world of the knower exists only as presented by *śabda* or as the meaning of *śabda*. Consequently, even conventionally unreal things like a fire circle (*alātacakra*), celestial city (*gandharvanagara*) and such others are cognized when presented by *śabda*.[25]

---

[22]*na so'sti pratyayo loke yaḥ śabdānugamād ṛte /*
anuviddham iva jñānam sarvam śabdena bhāsate // VP. I.123.

[23]K.A.S. IYER, *Bhartṛhari*, 100.

[24]*śabdeṣvevāśritaḥ śaktiṛ viśvasyāsya nibandhinī /*
yannetraḥ pratibhātmāyaṃ bhedarūpaḥ pratiyate // VP. I.118.

[25]*atyantam atathābhūte nimitte śrutyupāśrayāt /*
dṛśyate'lātacakrādau vastvākāranirupaṇā // VP. I.130.

The cognitions of babies are also fused with *śabda*. Babies are able to understand and react to other's speech because they are born with *śabdabhāvanā* or linguistic predispositions.[26] Hence, the knowledge of what is to be done and the inclination to express oneself depend on *śabda*. So for Bhartṛhari all activity of consciousness has *śabda* as its medium. If knowledge were to cease to be of the nature of *śabda*, consciousness would remain unmanifested to one's own self.[27] For Bhartṛhari, the speech-potency (*śabda-bhāvanā*) is the cause, of which the effect is audible speech. This audible speech is what we generally call verbal language (*bhāṣā*). The mute person possesses the former, not the latter. Hence his sensory awareness is not pre-linguistic. In conception-free awareness of adults the speech-potency is latent.

Śabda in Bhartṛhari's view pervades not only all perceptual knowledge but also inferential judgements. The intellectual powers which human beings possess are ultimately the functions of *śabda.* The ability to reason with others or transmit thoughts is based on the power of *śabda.*[28] Whatever be the metaphysical presupposition, its establishment and comprehension is based on *śabda.*[29]

4. *ŚABDATATTVA*: THE NON-DUAL

Most of the Indian scholars deal with *Vākyapadīya*, the master-piece of Bhartṛhari, as a work which has a non-dualistic

---

[26] *ādyaḥ karaṇavinyāsaḥ prāṇasyordhvaṃ samīraṇaṃ / sthānānāṃ abhighātaś ca na vinā śabdabhāvanāt //* VP. I.122.

[27] *vāgrūpatā ced utkrāmed avabodhasya śāśvatī / na prakāśaḥ prakāśeta sā hi pratyavamarśinī //* VP. I.124.

[28] *śabdānām eva sā śaktis tarko yaḥ puruṣāśrayaḥ / śabdān anugato nyāyo 'nāgameṣu anibandhanaḥ //* VP. I.138.

[29] *svamātrā paramātrā vā śrutyā prakṛmyate yathā / tathaiva rūḍhatām eti tathā hyartho vidhīyate //* VP. I.129.

philosophy as its foundation. As Pandey observes Bhartṛhari is regarded as an Advaitin by various authors and their works.[30] Murti views the School of Grammar as an *Advaitavāda* (Absolutism) which shares in the virtues and difficulties of all absolutism.[31] Manickam sees Bhartṛhari's philosophy as a sort of *Śabdadvaita* which forms the epistemological structure of Śaṇkara's Advaita hermeneutics.[32] According to Chakravarti Bhartṛhari was gifted with a true vedantic insight and has made use of such technical vedantic terms as *upādhi* (limiting adjunct) and *adhyasa* (illusion), to explain the diversity of words in the unity of sentence.[33]

Scholars also point to the fact that Bhartṛhari was influenced by the Madhyamika and *Vijñānavāda* discussions of the Buddhist traditions of his time about the nature of the ultimate Reality.[34] Moreover, he was thorough with the Vedānta conception of Brahman as the ultimate Reality underlying the whole universe.[35] This Brahman is somehow transformed into all things we experience in the universe.[36]

In the *Dravyasamuddeśa* (chapter on Substance) of the third *kāṇḍa* of *Vākyapadīya*, while discussing the nature of Brahman,

---

[30]S.L. PANDEY, *Pre-Śaṇkara Advaita Philosophy* (Allahabad: Darshan Peeth 1974), 334. Hereafter S.L. PANDEY, *Pre-Śaṇkara Advaita Philosophy.*

[31]T.R.V. MURTI, "Some Comments on the Philosophy of Language in the Indian Context," *Journal of Indian Philosophy* 2 (1974), 328.

[32]T. MANICKAM, "Towards an Indian Hermeneutics of the Bible," *Jeevadhara* 12 (1968), 103.

[33]P.C. CHAKRAVARTI, *The Linguistic Speculations of the Hindus* (Calcutta: University of Calcutta 1983), 109.

[34]H. NAKAMURA, "The Concept of Brahman in Bhartṛhari's Philosophy," *The Journal of Oriental Research* 40/41 (1981), 145.

[35]Br. Up. 1.4.10.

[36]Br. Up. 1.2.36.

Bhartṛhari interprets it as *dravya* (substance).[37] Bhartṛhari considered particular things as real or substantial in the everyday world but they are not real or substantial in the ultimate sense. For him, substance is the universal in particular things which is nothing other than Brahman as existence. Brahman as substance is also called *tattva* (thatness) which is further referred to as *para* (the highest one).[38] As *tattva* it neither is nor is not; it is neither one nor differentiated; neither combined nor separated; neither changing nor unchanging.[39] This eternal existent is made known to us through nonexistent forms (*ākāra*); likewise, it is this existent alone that is designated by linguistic forms that designate nonexistent limitations (*upādhis)*. To be different is limiting adjunct of the highest one and the same is true of not being different. It points to the fact that there exists no distinction between substance *(tattva*) and what is not substance (*atattva*). It would then seem that, the one Being is the universal substance; it is different from the finite substances *qua* substance in that the finite substances have qualities, whereas the one Being has no distinguishable quality; furthermore, of a finite substance, existence appears to be a predicate, whereas this one Being is the existence of which all finite substantial contents are, in reality, predicates.

Because the forms (*ākāra*) are essentially identical with

---

[37]The term *dravya* 'substance' in Bhartṛhari's system has a very special metaphysical sense. Bhartṛhari discusses this metaphysical sense of *dravya* in the second chapter of the third book of *Vākyapadīya*. As Helārāja, commenting on the third book, points out, the term *dravya* is used in two senses: (a) the ultimate reality or the ultimate substance - *pāramārthika dravya* and (b) the empirical substance - *sāmvyavhārika dravya*. (Helārāja's commentary on VP. III.2. (Poona: Deccan College Building Centenary and Silver Jubilee Series 71, 1971). The notion of empirical substance has been discussed by Bhartṛhari in the fourth chapter of the third *kāṇḍa*.

[38]VP. III.1.20.

[39]VP. III.2.12.

substance (whose forms they are supposed to be), language that designates them designates the eternal substance.[40] That (real substance), the form of which is not subject to conceptual construction (*avikalpita*), appears as if subject to constructions. For example, though not subject to temporal distinctions it appears as if it is. What is existent is that which remains at the end when forms disappear. It is eternal, expressed by language and not different from the language principle. It is neither existent nor nonexistent; neither single nor separate; neither connected nor disjoined; neither modified nor unmodified. This single (substance) is seen as language, meaning, and the relation between them. It is what is seen, seeing, the seer, and the result of seeing.[41]

As prof. Nakamura observes, since the Absolute transcends all spatial and temporal determinations, differentiations based on concepts like identity, difference, being and non-being are not applicable to it. For these reasons diversity and multiplicity could not really be part of the essence of the Absolute.[42] In the same way the various forms of changes can in no way be established in true Reality. The Reality seems to have these forms as its essence but it is something whose nature is not at all comprised of these forms.[43]

Bhartṛhari declares in unambiguous terms that the ultimate Reality is a single unitary entity. He says that it is the one ultimate which differentiates on account of its different powers. Existence and non-existence are two appearances and are not apart from the

---

[40]VP. III.2.6.

[41]H.G. COWARD & K. RAJA, ed., *Encyclopedia of Indian Philosophies* Vol. V, *The Philosophy of the Grammarians* (Delhi: Motilal Banarsidass 1990), 157.

[42]H. NAKAMURA, "The Concept of Brahman in Bhartṛhari's Philosophy", 157.

[43]VP. III.2.10.

one self.[44] In Bhattacharya's view Bhartṛhari's approach to the phenomenal entity is essentially the same as that of Śaṇkara's Advaita. For Śaṇkara every phenomenal entity though false as an appearance, has yet a core of reality, hidden under its outward sheath, which consists of *nāma-rūpa* (name and form).[45]

This means that the entire world of things whose individuality consists only in names and forms (*nāma and rūpa*) to use the advaitic terms, has its source in this speech-essence. It does not mean that the *Śabdatattva,* the eternal timeless, is limited by time. Instead, time is an inherent power of the Absolute which exerts its influence in bringing about the powers of *Śabdatattva.* So, Bhartṛhari, explains all the divergent manifestations such as individual subjects, the objects and all the relations on the basis of the various powers of Brahman.

According to Dasgupta: "The difference of the world of appearances consists in the fact that for the time being it appears to have its nature as brahman completely subordinated, and instead of having itself manifested as brahman it appears only as material objects."[46] He says further that Bhartṛhari was quite willing to accept the ordinary realistic view of appearances (which Śaṇkara refuses to do), but from the ultimate point of view Bhartṛhari would not for a moment admit the existence of any thing else but the Ultimate Being.[47]

Above discussions show that Bhartṛhari looks upon phenomenal creation, as made up of two aspects- one real and the other unreal. The essence of the real is pure Being (*sattā*), which is manifested as *jāti* (universl) in all particulars (*vyaki*). It is *jāti*

---

[44]VP. III.3.61.

[45]B. BHATTACHARYA, *Bhartṛhari's Vākyapadīya and Linguistic Monism* (Poona: Bhandarkar Oriental Research Institute 1985), 18.

[46]S. DASGUPTA, *Indian Idealism* (London: Cambridge University Press 1962), 196. Hereafter S. DASGUPTA, *Indian Idealism.*

[47]S. DASGUPTA, *Indian Idealism*, 196.

in the form of *sattā* that is the noumenon, while the *vyaktis* are mere appearances, having varying names and forms. He elucidates this point further in the *Jātisamuddeśa* (chapter on Universal Property) of the third *kāṇḍa* of *Vākyapadīya*. The individual is non-existent (*asatya*) and it is the universal that is the existent (*satya*). Indeed, it is the existence (*sattā*) alone that differentiated through its relata is said to be the universal and all the language is based on that existence. Hence it is the universal that is the prompter (*prayojaka*) of the coming to be of the particular, being present antecedently in its causal conditions.[48] Even the three-fold differentiation of the Absolute into *bhoktṛ* (enjoyer) *bhoktavya* (enjoyed) and *bhoga* (enjoyment) has its validity only in the plane of our mundane existence.

As has been pointed out already reality manifests itself as many because of its many powers or potencies. These powers can be divided into two categories as direct and indirect powers of Brahman. Time, space, direction, action, generally come under direct powers. Inherence, instrument, sequence, permission, suspension, decomposition, present, past, future, etc. come under indirect powers.[49]The *Sādhanasamuddeśa* (On the Means to Action) section of the third *kāṇḍa* recognizes these potencies with a view to explaining the workings of the world of appearance in which we live and have our being.[50] Bhartṛhari believes that all actions are transformations (*vivarta*) of Being (*bhāva)* which takes place in the sixfold forms of birth (*jāyate)* existence (*asti*) change (*viparivartate*), growth (*vardhate*), decay (*apakṣīyate*) and destruction (*vinaśyati*).[51] This manifestation does not mean that Reality loses its oneness. Hence the supreme word of

---

[48]VP. III.I.32-34.

[49]P.S. SHARMA, "The Idea of Power (*śakti*) in the *Vākyapadīya*," *Bharatiya Vidya* 41 (1981) 75-79.

[50]VP. III.7.

[51]K.A.S. IYER, *Bhartṛhari*, 247.

Bhartṛhari is a monistic principle and the phenomenality is only a *vikalpa* (logical construction) serving as the explanation for phenomenal plurality.

Although the Absolute is one, Bhartṛhari considers the differenciated aspects of the phenomenal world as based upon it. This would mean that the Absolute has some form of connection with the differentiated aspects. Thus Bhartṛhari says: "That does not exist and yet exists, that is one and yet different, that unites and separates, that changes and is different from it (i.e., does not change)."[52] This again is an indication to the transcendental and immanent aspects of Reality. Bhartṛhari compares both aspects and further states: "(In the highest truth) the truth itself which is not discriminated receives the forms of discrimination. And differentiation in time does not exist in that truth, yet differentiation in time is perceived."[53]

One is Brahman in the sense of highest truth, and the other is the aspect which forms the differentiated aspects of the phenomenal world. If the source which unfolds the differentiated aspects of the phenomenal world is in the Brahman itself, the essence of the Brahman in ultimate sense must be plural, because the potentiality of differentiation resides within Brahman. If such is the case, the idea that the highest Brahman is an Absolute one is contradicted. To solve this difficulty was the central concern of all the Vedānta scholars. In contrast to the other philosophers especially the Vedantins, the Grammarian emphasized the fact that Brahman as the basis of differentiated aspects is of the nature of *śabda* which he called *Śabdatattva*. It is the greatest characteristic of Bhartṛhari's philosophy which yearned him the name of *Śabdabrahmavādin* or *Śabdādvaitavādin*.

According to E.R.S. Sarma, Bhartṛhari was an Advaitin in his

---

[52]VP. III.2.13.

[53]VP. III.2.8.

metaphysics. Further he points out that Helārāja, the commenteter of the third part of *Vākyapadīya* presents Bhartṛhari's views as nothing but Advaita.[54] However, the Advaitins in general have difficulty with Bhartṛhari's contention that Brahman is *Śabdatattva*. Nevertheless, the Vedāntic passages quoted in the *Vṛtti* to show identity between *vāk* or *oṃkāra* and Brahman are supportive to the contention of Bhartṛhari.[55] But Śaṇkara does not interpret them in their primary sense; he takes them in the secondary sense.[56] For example, the fisr verse of *Māṇḍūkya Om iti etad akṣaram idam sarvam*, would, according to Śankara mean that *Oṃkāra* is the sound indicating Brahman, whereas Bhartṛhari holds Brahman itself is the word-essence. In later times Kumārila (c.660) of the Mīmāṃsā School, Jayanta Bhaṭṭa (840-900) of the Nyāya School and Vimuktātman (1200) of the Vedāntists School have attempted to refute Bhartṛhari's *Śabdatattva* concept on various grounds.

However, Maṇḍana Miśra, (c.690) one of the early advaitic thinkers, elucidates the *Śabdādvaita* view of the Grammarian in his celebrated work *Brahmasiddhi*. As Potter points out Maṇḍana's thesis appears to be that Brahman is consciousness, that consciousness is the power of speech, and so Brahman is of the nature of speech; the whole universe is a manifestation (*vivarta*) of speech.[57] Explicating further the sense of the term

---

[54]E.R.S. SARMA, "Bhartṛhari as an Advaitin," *Adyar Library Bulletin* 49 (1986) 579.

[55]VPV. I.1; I.4; I.45.

[56]For Śaṇkara's refutation of the Grammarians view, see M.F. MÜLLER, ed., *Sacred Books of the East* (Delhi: Motilal Banarsidass 1973), Vol. XXXIV: *Vedānta Sūtras with the Commentary by Śaṇkarācārya*, Part I, tr., by G. THIBAUT, 1.3.24 - 1.3.39.

[57]K.H. POTTER, *Encyclopedia of Indian Philosophies*, Vol. III, *Advaita Vedānta up to Śaṃkara and His Pupils* (Delhi: Motilal Banarsidass 1981), 54.

"akṣaram" he presents a supportive view of the *Śabdatattva* concept of Bhartṛhari. In contrast to Śaṇkara, Maṇḍana takes the word *"akṣaram"* to mean that Brahman is of the nature of *śabda*. In his opinion the upanishadic texts like *Om iti Brahma, Om iti sarvam*[58] point to the identity between Brahman and the sacred syllable *Om* which is also called *Praṇava*.

5. *SPHOṬA*: THE GROUND OF MEANING

Bhartṛhari's linguistic theory of the *sphoṭa* also forms part of his monistic and idealistic metaphysical theory according to which the transcendental speech essence is the first principle of the universe. This *sphoṭa* theory, in brief, says that the word essence is the very soul of a meaningful word. It is something that the word reveals or communicates to mind through its meaning. So what is eternal and self subsistent is not the sound or the word, but the word-essence. The word-essence which is the Absolute for Bhartṛhari appears on the one hand, as the plurality of words and, on the other hand, as the plurality of objects. Hence, the Word-Absolute is ultimately one and undivided. Yet in our empirical experience it appears as divided into sentences, words and phonemes.

The Grammarian holds the sentence to be an indivisible unit (*vākya sphoṭa*) of meaning. The division of sentence (*vākya*) into terms (*pada*) according to Bhartṛhari, is an unreal abstraction. Such a division may be necessary in order to make the unintelligent understand the nature of a sentence but considered in themselves the terms are nothing but fictions, pure and simple. Though the division is fictitious, it is useful in understanding the indivisible sentence.[59] Just as the indivisible sentence *sphoṭa* does not really consists of any parts, so the meaning denoted by it does

---

[58]Tait. Up. 1.8.

[59]VP. II.7-10.

not admit of any division.[60] Both the form of, and the object denoted by, a word conform to the same principle.

Again, the Grammarian recognizes one ultimate *sphoṭa*, which is infinite, eternal and self-existent. Each particular *sphoṭa* latent in a word or a sentence is but an apparent variation of the universal, immutable *sphoṭa*. So *sphoṭa* is essentially one and is identical with Brahman. That is why the advocates of the *sphoṭa* theory conceive imperishable Brahman, without beginning and without end as the eternal principle of word (*Śabdatattva*), which means the essence of all letters, words, sentences and their meanings. Emphasizing the transcendental oneness and the phenomenal diversity of *sphoṭa*, Murti says thus:

> In linguistic apprehension, as in other cognitions, there is the interplay of two factors of different levels- the empirical manifold of sense data (the separate letters of words in this case) and the transcendental or a priori synthesis of the manifold which alone imparts a unity to those elements which would otherwise have remained a mere manifold.[61]

6. *ARTHA*: MEANING OF THE REAL

Bhartṛhari's discussions on the relation between word and meaning also have their bearing on his metaphysical outlook. For Bhartṛhari, as has been observed elsewhere, the relation in which speech and thought have mingled together is inseparable. Hence *śabda* and *artha* are but different aspects of one and the same thing.[62] What one is asked to aim at is a knowledge of this undivided character of *śabda* and *artha*. As Chakravarti views, comprehension of the undivided character of *śabda* and *artha*

---

[60] *śabdasya na vibhāgo 'sti kuto 'rthasya bhaviṣyati /*
vibhāgaiḥ prakriyābhedam avidvān pratipadyate // VP. II.13.

[61] T.R.V. MURTI, "Some Comments on the Philosophy of Language in the Indian context," 329.

[62] *arthabhāgaiḥ tathā teṣām āntaro 'rthaḥ prakāśate /*
ekasyaiva ātmano bhedau śabdārthau apṛthaksthitau // VP. II.31.

takes place only in a stage of higher spiritual cultivation when all forms of dualism disappear or merge into the unity of the Infinite.[63] Even after their undivided nature word and meaning appear to be distinct from each other only through the intervention of *māyā* (illusion).[64]

In this connection Bhartṛhari goes on to say without any reservation that knowledge is translated into word. Words that we speak come out as an embodiment of thought. In the process of transformation ideas which are in the intellectual form become vivid and precisely complete in structure when they are expressed through words. The internal knowledge, holds Bhartṛhari, remaining within as a subtle element of speech, assumes the concrete form of words when necessity arises for the purpose of revealing its own identity.[65] What we understand from this assertion is that the words as bearers of meaning are eternal and they are eternally and inseparably related to the Absolute Existence. And for Bhartṛhari it is natural and not created.

Words by their very nature refer to objects. If the objects are knowable they cannot remain without words, for no knowledge is possible without words, every knowledge is in the form of language, says Bhartṛhari.[66] It means that the knowledge content comes from its root epistemological cause, i.e., the *vāk*; word. The *Vākyapadīya* I.4 makes it clear thus: that the experiencing subject, the object of experience and the process of experiencing are all contained with the supreme Word. It is only thought as expressed in word that can be understood and communicated. All these observations definitely point to the absolutistic nature of Bhartṛhari's thinking.

---

[63]P.C. CHAKRAVARTI, *The Linguistic Speculations of the Hindus,* 325.

[64]P.C. CHAKRAVARTI, *The Linguistic Speculations of the Hindus,* 336-337.

[65]VP. I.113.

[66]*na so 'sti pratyayo loke yaḥ śabdānugamād ṛte /*
anuviddham iva jñānam sarvam śabdena bhāsate // VP. I.123.

7. THE REAL AND THE TIME

To explain the emergence of the world of immense diversity out of a single unity, the Grammarians felt obliged to recognize the various powers of the ultimate principle. Bhartṛhari has tried to explain the plurality by having recourse to the concept of *śakti* (power) inherent in the Word-Principle. Thus in the very second *kārikā* of *Vākyapadīya,* Bhartṛhari presents the One as having no second. It manifests itself as distinct and manifold, having an independent status as it were, through the working of various powers. These powers or potencies are of diverse sorts, some of which are direct and others indirect, and though they are identical with *Śabdatattva*, they seem to be different from it. In truth, *Śabdatattva* is a totality of potencies which have no separate independent being of their own apart from their substratum which is One, Indivisible and Eternal.[67] As to the relation of these powers with the source, the *Śabdatattva*, Bhartṛhari declared that all the powers are identical with *Śabdatattva*.[68] Bhartṛhari treats the *kālaśakti* (time power) as the supreme of all the powers of the Absolute because it controls all other powers of *Śabdatattva* by presenting them in a particular temporal order.

The creation of the objects of the universe occurs in the first instance through the creative power of Brahman's *Kāla* or Time-power. *Kāla* is not different from Brahman but is that aspect of Brahman which follows manifested sequence to come into being. When such time sequences appear as differentiated objects then time as a power seems to be different from Brahman, but really it is not. Through time things come to be and through time things pass away. Time is the efficient cause by which Brahman controls the cycles of the universe. In a separate section on time called *Kālasamuddeśa*, Bhartṛhari illustrates the creative process of time

---

[67]B. BHATTACARYA, *Bhartṛhari's Vākyapadīya and Linguistic Monism*, 11-12.

[68]VP. III.1.22.

like that of a wire-puller in a puppet-play.[69] Just as the wire puller is in complete control of the puppet-play so *Kāla* has full control over the running of the world. Ordinary cause and effect processes cannot operate unless time-power infuses them with life-force.

Again in the third verse of the *Kālasamuddeśa* time is said to be the cause of activities like creation, existence and destruction of beings. Time permits (*abhyanujñā*) some things to appear at a particular time and prevents (*pratibandha*) others from appearing. On the one hand by its permissive function, time allows things to be born and to continue in existence[70] and on the other time obstructs the inherent capacities of objects with its preventive function and *Jarā* (old age) is then experienced. Time which is one due to its activity of growth and decay attains the states of past, present, and future. Thus when an action ceases, time conditioned by that action is called past. When something is about to happen, time, conditioned by that event is called future. When action has been initiated but is not yet completed, time is then called present.[71] In this way time is experienced through the actions of the secondary causes it releases or restrains, to be sequenced into past, present, and future. The appearance of the universe, which is really without sequence, as something with sequence is the work of time.[72]

If Real is one as Word in the system of Bhartṛhari, how does it manifests itself as many even though one, as occurring in sequence even though by itself beyond all sequence and ever-existent, as arranged in time and space even though itself non-temporal and non-spatial in character, as subject and object, as

---

[69]BHARTṚHARI, *Vākyapadīya* Chapter III. Part. II, tr., K.A.S. IYER, (Delhi: Motilal Banarsidass 1974). III.9.4.

[70]VP. III A.9.24.

[71]VP. III A.9.37.

[72]VP. III A.9.46.

cogniser and the thing cognised, as the expressive word and the meaning expressed thereby and similar other hypostatizations, even though by itself it is indivisible, is a vital question. The Vedānta as interpreted by Śaṇkara has sought the help of *māyā* for explanation of plurality and other appearances in the world of phenomena. Bhartṛhari has tried to explain it by having recourse to the concept of *śakti*. According to Coward the description of time in Bhartṛhari's system sounds very similar to Śaṇkara's notion of *māyā* in relation to Brahman.[73]

According to the Vaiśeṣika system of philosophy, time, space, direction, inherence, action, etc. are substances and different categories (*padārtha*). They accept time as an eternal substance uncaused by anything. However they do not consider time as a power.[74] But Bhartṛhari calls them all powers. For him the entire universe is a bundle of powers. The Brahman, according to him, possesses unlimited powers and as a matter of fact the Brahman is not different from the powers or the power inherent in *Śabdatattva*. In the following *kārikā* of *Vākyapadīya*, Bhartṛhari declares:

> *śabdeṣvevāśritā śaktiṛ viśvasyāsya nibandhinī /*
> *yan netraḥ pratibhātmāyaṃ bhedarūpaḥ pratiyate //*
>
> The power which is based on words controls this universe. This universe which has a single Intelligence as its soul is perceived as manifold through the word as the eye.[75]

Though the *Śabdatattva* is one without a second, it manifests itself as distinct and manifold, having an independent status as it were, through the working of various powers or potencies inherent in itself. These potencies are of diverse sorts, some of which are direct and others indirect, and though they are identical

---

[73]H.G. COWARD, "Time (*kāla*) in Bhartṛhari's *Vākyapadīya*," *Journal of Indian Philosophy* 10 (1982) 279.

[74]P.S. SHARMA, *The kālasamuddeśa of Bhartṛhari's Vākyapadīya*, 31.

[75]VP. I.118.

with *Śabdatattva*, they seem to be different from it. For Bhartṛhari, Brahman, as the Word-Principle, is an intrinsically dynamic and expressive reality and the universe as a whole is to be understood as its manifestation under the form of temporal becoming. This Brahman, the *Śabdatattva* (Word-Principle) is not a lesser Brahman, a mere *upāyaḥ* (means), but is identical with Brahman itself, the only Brahman there is.[76]

8. REAL AND THE PROBLEM OF CHANGE

Brahman possesses many powers due to the operation of which it appears as the differentiated world. Of these, time (*kāla*) and space (*dik*) are foremost in projecting the infinite as the manifold but finite world. The time-power refers to the capacity of Brahman to present itself in a sequence. The space-power presents reality as limited and possessing dimensions. Though not different from its powers, Brahman appears to be so. The powers are not identical with Brahman because such an identity would bring about duality in the non-dual Brahman. They are not different as they cannot exist apart from Brahman. As they do not have an existence independent of Brahman, they are also not real. Yet, their effects are seen in the world, so they are not unreal either. They are not mutually contradictory since they exist simultaneously in the same substratum, namely, Brahman. But as they give rise to different effects, they are distinct from each other. In any of their aspects the powers do not affect the non-dual nature of Brahman.

This non-dual Brahman which is the source of everything, appears also as the enjoyer, enjoyed and enjoyment. It is the subjective as well as the objective world and the relation between

---

[76]M. BIARDEAU, *Theorie de la connaissance et philosophie de la parole dans le brahmanisme classique* (Paris: Mouton 1964), 266ff.; see also H.G. COWARD, *Derrida and Indian Philosophy* (NewYork: State University Press 1990), 37.

the two. Bhartṛhari refers to the view of the knowers of the Vedānta who uphold the reality of the non-dual self in whom the different functions of knower, known and knowledge are conceived.[77] They hold that the cognition of differentiation is due to the ignorance of the true nature of Reality. Brahman the formless, assumes forms, when conditioned by limited objects which are in fact, effects of *avidyā* (nescience). In relation to these objects and the empirically limited knowledge, Brahman itself is viewed as the knower. Like the Vedāntins, Bhartṛhari also considers that the ultimate reality appears divided into the knower, the object known and the knowledge as well as the purpose of knowledge.[78] Brahman is everything that is dynamic and static in the world, for, it is action (*kriyā*) when its time-power is active and substance (*sattā*) when sequence ceases to operate.[79]

These transformations however, do not affect Brahman. The all-pervasive, timeless and non-relational reality appears as the world bound by space, time and causality which does not affect Brahman's integrity. Bhartṛhari upholds the doctrine of apparent change (*vivarta*); thus Brahman retains intact its pure state of absoluteness. The manifestation of the world is thus nothing but an unfolding of Brahman's self nature as different from itself. Since all that exists is in fact the un-differentiated one, there can be neither origination nor destruction. According to Bhartṛhari, all differentiation, all causation and change is but apparent. Although he does not specifically discuss the concept of causality in the context of Brahman and its transformations, he shows in the *Sambandhasamuddeśa* (chapter on Relation) of the third *kāṇḍa* that the concept of causality and the notions of existence and non-

---

[77]VP. III.3.72.

[78]VP. III.2.14.

[79]VP. III.1.35.

existence which form the basis of this concept are unsatisfactory.

Bhartṛhari states that the notions of existence and non-existence are mutually dependent and lack absolute reality because of their interdependent nature. They are interdependent because one cannot speak of non-existence without reference to existence and vice versa. Even if existence and non-existence are independent entities, they are still unintelligible because non-existence cannot become existence nor existence change into non-existence. This is because non-existence being inexplicable, a cause cannot work towards it and what is already existent does not need any cause to bring it into being.[80] However, existence and non-existence cannot be entirely different from each other. If they were unrelated, non-existence would not remove existence nor existence be produced on the destruction of non-existence.[81]

While accepting the conception of existence and non-existence as essential for empirical activity, Bhartṛhari considers them as appearances from the absolute point of view.[82] These two states are not wholly different from the non-dual self on which they are superimposed. Non-existence is nothing but a state of imperceptiblity. An object is held to be non-existent when it is conditioned by the states of past and future. But it is believed to be present when it is delimited by the present time and is cognized.[83] Thus the two states of existence and non-existent are adventitious to the Ineffable.[84]

Bhartṛhari seems to reject both doctrines of *satkāryavāda* (i.e., effect is existent in a subtle form in its cause) and *asatkāryavāda* (i.e., effect non-existent prior to manifestation). In his view prior to the coming into being an effect is held to exist in mind, i.e., it

---

80VP. III.3.60-62.

81VP. III.3.74-75.

82VP. III.3.71.

83VP. III A.9.49.

84VP. III.3.61.

comes to possess a referential existence (*upacārasattā*) or secondary being and is believed to be the subject of the act of being born[85]. Hence the effect is neither existent nor non-existent but remains as a mental image possessing a secondary Being. The object in the mind is held to be the effect of the action of creation.[86]

The change which causation brings on is only apparent because in essence the effect does not differ from the cause. But at the end Bhartṛhari concludes that the appearance of objects is due to an extremely wonderful mode (*atyadbhutavṛtti*).[87] The real essence of all manifestation which is in fact devoid of division and sequence seems to appear as individual entities possessing spatial and temporal limitations. Not only their creation but also their sustenance and destruction are through this miraculous activity.[88] Thus from the absolute point of view, Brahman cannot cause nor change itself into the phenomenological world for causality is self-contrdictory and change is illogical.

Since Bhartṛhari rejects causation, the question arises whether he also dismisses the world as unreal. Closely related to this issue is the question whether Bhartṛhari accepts the concept of nescience (*avidyā*) as the root cause of the appearance of the world and if he does, then what exactly is the relation between Brahman and the phenomenal world. The word *avidyā* is mentioned only twice in the *kārikās*. It occurs for the first time in the *Vākyapadīya* II.233, and is used in a linguistic context. Bhartṛhari says that it is *avidyā* that is described in the science of grammar. *Vidyā* which signifies unity of sentence and meaning arises independent of technical learning. Though according to the Grammarinas the sentence is an indivisible unity, they divide it

---

[85]VP. III.3.45.

[86]VP. III.7.105.

[87]VP. III.3.81.

[88]VP. III A.9.26.

into words which are again sub-divided into roots suffixes, etc. Bhartṛhari refers to such fictitious differentiation of the word as *avidyā*..

The second reference to *avidyā* is more metaphysical in nature. Bhartṛhari states that time power is the first to appear in *avidyā* and that it does not exist in *vidyā*.[89] Both verses seem to denote by the term *avidyā* a state of multiplicity which is not ultimately real and which is removed when the true knowledge of the non-dual undifferentiated reality (*vidyā*) arises. Again *avidyā* is referred to as a power which projects Brahman as the differentiated world. Its relation with the other powers is not clear from these references. Nevertheless, it is possible that the term *avidyā* is being made use by Bhartṛhari to denote all the powers of Brahman.

9. PHENOMENALIZATION OF THE REAL

The question whether Bhartṛhari is a *vivartavādin* who holds the world to be an unreal appearance or a *pariṇāmavādin* who upholds the creation as a real transformation has been a subject of controversy among scholars. Ancient Indian Philosophers like Vācaspatimiśra, (900-980) Jayanta Bhaṭṭa and others refer to the two-fold explanation (namely, *pariṇāmavāda* and *vivartavāda*) of the philosophy of Bhartṛhari. At the same time scholars like Bhaṭṭoji Dīkṣita, (c.1590) Prabhākara (c.700) and others refer to the *vivartavāda* of the Grammarians. G. Sastri has maintained that Bhartṛhari was not familiar with the difference in the connotations of the terms, *pariṇāma* and *vivarta*.[90] He bases his view in the following *kārikā*, where Bhartṛhari has used the two terms *pariṇāma* and *vivarta* as synonyms. The *kārikā* goes thus: knowers of traditions (*Āgamas*) have declared that all this is the

---

[89]VP. III A.9.62.

[90]G. SASTRI, *Philosophy of Word and Meaning*, 15.

transformation (*vivarta*)of the word. It is from the *śabda* that this universe has evolved (*pariṇāma*).[91]

While providing an explanation for the dual interpretation of *Śabdadvaita*, Pandey asserts the existence of two schools of philosophy within the Grammar schools, i.e., the school of Patañjali and the school of Bhartṛhari. The former is realism or *pariṇāmavāda* while the latter is idealism or *vivartavāda*.[92] In the hands of Bhartṛhari, the status of the phenomenal world in relation to *Śabdatattva* cannot be adequately characterized by either *pariṇāma* or *vivarta* in their respective doctrinal formulations. Bhartṛhari does not seem to believe in any real change of *Śabdatattva*, therefore, conception of *pariṇāma* is ruled out. Nor does the *Vākyapadīya* seem to contain exactly the same kind of logical antimony between the ground which is an indivisible unity and its appearances, which forms the basis for the *vivarta* concept.

However, the phenomenalization from Bhartṛhari's view point attracted much attention of later interpretators and critics. Consequently there appeared divergent interpretations of Bhartṛhari's *Śabdadvaita*. Among them two different interpretations gained momentum during the course of time. One of which looks upon the phenomenal world as real as in the Sāṃkyan sense of *pariṇāma* (a real transformation) and the other looks upon the phenomenal world as something unreal, in the Advaitic sense of *vivarta* (an unreal appearance).

According to G. Sastri the followers of Bhartṛhari came to interpret *Vākyapadīya* in the light of *vivarta* influenced by the advaitic use of the term. The *Vṛtti* on the first *Kāṇḍa* (which he considers as that of Puṇyaraja,) is a later commentary and Bhartṛhari cannot be held to be a *vivartavādin* on the basis of the

---

[91]VP. I.120.

[92]S.L. PANDEY, *Pre-Śaṇkara Advaita Philosophy*, 343.

reference in it. He states further that unlike the *avidyā* of *vivartavādins* the powers conceived by Bhartṛhari are as real as Brahman. He remarks that there is no clear assertion of Bhartṛhari that the world is an illusion.[93] He notes further: "unlike Śaṃkara, he (Bhartṛhari) views the Supreme Reality not as a mere locus or a passive background for the play of the power (*śakti*) but as power itself. Freedom to act is only natural to the Eternal Verbum."[94]

Further, Sastri does not find any difference between *Śabdabrahman* and its *śaktis* in the system of Bhartṛhari. He claims, Being (*Śabdabrahman*) and its powers (*śakti*) or appearances are identical. Therefore, plurality of powers in Bhartṛhari's system has no independent being apart from the Absolute Word and each member of the plurality partakes of the character of the word, as the powers are inherent in and identical with the Absolute. Whereas in Śaṇkara's Vedānta the unity is absolute and unqualified and plurality of powers is a mere appearance (*māyā*) which has no ultimacy.[95]

In Bhartṛhari's system the powers of the Absolute, like time, space, etc. are ultimately real and have the same ontological status as the Absolute itself which do not imply dualism. But Śaṇkara would never accept the ultimate Reality of any power (*śakti*) co-existing with the Absolute. In Śaṇkara's system the power of the Absolute is called *māyā* which is attached to it in a most mysterious way. *Māyā* is neither identical with nor distinct from the Absolute. Since *māyā* is not the essence of the Absolute it is regarded as a false adjunct. According to Sastri's observation, none of the stanzas from *Vākyapadīya* speaks about the powers of the Word-Principle as unreal. So he takes it for granted that for

---

[93]G. SASTRI, *Philosophy of Word and Meaning,* 64.

[94]G. SASTRI, *Philosophy of Word and Meaning,* 19.

[95]G. SASTRI, *Philosophy of Word and Meaning,* 20.

Bhartṛhari, the phenomenal world which is a creation of powers is real and not a *vivarta*, a mere illusory appearance. He maintains that many of Bhartṛhari's commentators have failed to realize the above distinctions because they have read the *Vākyapadīya* under the influence of later vedāntic *vivarta* formulations.

Again Sastri observes the Absolute of Bhartṛhari as dynamic principle which produces the universe out of itself. It appears to be the material and the efficient cause of all that exists. What Sastri concludes from these observations is that the Absolute of Bhartṛhari is endowed with multiple powers which are as real as the Absolute. According to Sastri's contention Bhartṛhari's position is much closer to the notion of 'reflection' (*ābhāsa*) formulated by the Trika writers of Kashmir. In this view the *śakti* and their phenomenal manifestation as words are identical with the Absolute. The relationship between the two described on the analogy of the mirror (the Absolute) and its reflection (the *ābhāsas*); the latter can have no independent existence without the former. Yet, the latter also have a reality which is somehow identical with the former.[96]

There are also scholars whose speculation go very much in line with that of Sastri. Among them M. Biardeau's observations are worth noticing. In her view, phenomenal world of Bhartṛhari's conception is real as it does not conceal the Being but reveals It. She indicates the logical necessity for the system of Bhartṛhari to postulate the Being as pure unity and non differentiation in order to explain the profound correlation of the objective and subjective aspects of reality. To quote:

> The phenomena do not veil the Being, but reveal it, or rather, the Being manifests itself through them ... The Being is not one and then many in the course of its evolution. Nor is it one really and many illusorily but it is one substantially and many phenomenally ...Thus one understands why it would be inconceivable to look for and, à

---

[96] G. SASTRI, *Philosophy of Word and Meaning,* 25.

> fortiori, to find in Bhartṛhari *avidyā*, metaphysical ignorance and *māyā*, cosmic illusion, which form the pivots of Vedantic thought.[97]

Hence she maintains that it is inconceivable to find in Bhartṛhari *avidyā*, metaphysical ignorance and *māyā*, cosmic illusion which form the pivots of vedāntic thought.

J.C. Plott compliments Biardeau's stand regarding Bhartṛhari's position, namely, realism. According to him, her view seems to be in direct opposition to the Buddhist theories before Bhartṛhari and the Advaitins after him which universally took the phrase *nāma-rūpa* (name and form) as identical with *saṃsāra* (phenomenal world) and therefore the realm of unreality. He contents Bhartṛhari's assertion as exactly opposite. For names and forms are real because they participate in Reality.[98] H. Ganguli's observation seems to go along with the same line of thought:

> Bhartṛhari has elaborately shown how existence and non-existence both are merely logical categories ... the difference between existence and non-existence is itself a fictional abstraction. ...Bhartṛhari has made the remarkable deduction that the issue between Realism and Idealism is ultimately a fictitious issue. The philosophers who assert that the world exists and the philosophers who assert that the world does not exist do not go for making any material difference between themselves, since both of them are dealing with fictions, because non-existence is as much a logical existent as existence itself.[99]

From these and other considerations Plott sees Bhartṛhari as a philosopher who belongs to the realist tradition of Indian philosophy. In his view Bhartṛhari's *Śabdavāda* is a realist

---

[97] M. BRAIDEU, *Theorie de la connaissance et philosophie de la parole dans la brahmanisme classique*, 309-310.

[98] J.C. PLOTT, *Global History of Philosophy*, Vol. III. (Delhi: Motilal Banarsidass 1980), 171. Hereafter J.C. PLOTT, *Global History of Philosophy*.

[99] H. GANGULI, *Philosophy of Logical Construction* (Calcutta: Sanskrit Pustak Bhandar 1963), 119. Hereafter H. GANGULI, *Philosophy of Logical Construction*.

alternative to idealist *Māyāvāda.*[100] He also perceives an Advaitic overlay on Bhartṛhari's philosophy. Plott overlooks Bhartṛhari's position somewhat like the Madhyamika "neutralism."[101] Further he identifies Bhartṛhari's Absolute with the *Saguṇa* Brahman (Brahman endowed with attributes) in Rāmānuja's system.[102]

One cannot deny the possibility that Bhartṛhari might have developed his vision in contrast to the Buddhist theories prevalent in his times, specially that of the Yogacāra School of Idealism, for whom the phenomenal world is a body of externalized concepts.[103] Bhartṛhari points real nature of names and forms in opposition to the Buddhist *saṃsāra*, for the obvious reason that they participate in the reality of meaning-conveyingness through the external expressions into the meaning content *sphoṭa* which is the *Śabdatattva*, the ultimate meaning of existence itself.

10. REAL TRANSCENDS THE PHENOMENAL

So far we have discussed certain interpretations of Bhartṛhari's philosophical stand point from a realistic perspective. Now, there are also scholars who find in Bhartṛhari an idealistic philosopher and interpret his thought accordingly. Bhartṛhari's work does not help us substantially, to trace Bhartṛhari's attitude to the doctrines of *pariṇāma* and *vivarta*. But Bhartṛhari's own commentary (*Vṛtti*) on the *Vākyapadīya* sheds much light into this problem and thus, the idealistic interpretations depend very much on the authority of the *Vṛtti*.

K.A.S. Iyer claims strong evidence for interpreting Bhartṛhari's thought in line with traditional Vedānta doctrine. He bases his interpretation on the assumption that Bhartṛhari is the author of the *Vṛtti* (as tradition maintains), although he admits

---

[100] J.C. PLOTT, *Global History of Philosophy*, 172.

[101] J.C. PLOTT, *Global History of Philosophy*, 170.

[102] J.C. PLOTT, *Global History of Philosophy*, 166.

[103] H. GANGULI, *Philosophy of Logical Construction*, 61.

that this question is not yet finally settled. In contrast to the views held by G. Sastri, M. Biardeau and others Iyer finds strong evidences for attributing to Bhartṛhari, a *vivarta* theory very close to that of Advaita Vedānta. In support to his arguments Iyer upholds Bhartṛhari's repeated description of the cognized phenomenal world as *asatya* (unreal) and his declaration that *satya* (real) is that which, although free from differentiation, assumes differentiation.[104] This is further supported by Bhartṛhari's repeated reference to *avidyā* as the cosmic power by which the one changeless reality is presented as having many word-forms arranged in a spatial and temporal sequence.

According to Iyer's observation while explaining the metaphysical position of Bhartṛhari, the oldest commentary *Vṛtti* and those of Helarāja and Puṇyarāja, bring in the principle of *avidyā* as the cause of multiplicity and differentiation in the one. The *Vṛtti* on the first *kārikā* declares the objects of the world unreal and cites from an ancient text which holds that everything which is conditioned by time and space is a result of the functioning of the power called *avidyā* and is undefinable both identically and non-identically with the Absolute (*vidyā*).[105]

As we have seen earlier, *Vākyapadīya* II.233 and III.9.62 do speak explicitly about *avidyā*. Iyer opines that in both contexts the word *avidyā* stands for a metaphysical principle, the division of the one into many, a division which in some way has a lesser reality than the undifferentiated one. Although the terms *pariṇāma* and *vivarta* appear more or less as synonyms in *Vākyapadīya* I.120 in the description of the emergence of the phenomenal world from the *Śabdabrahman,* Iyer concludes that all the available ancient commentaries, beginning with the *Vṛtti*,

---

[104]K.A.S. IYER, *Bhartṛhari*, 132.

[105]K.A.S. IYER, *Bhartṛhari*, 128.

interpret Bhartṛhari's thought as *vivartavāda.*[106]

As Chakravarti argues, in the very opening verse of the *Vākyapadīya* Bhartṛhari has referred to the doctrine of evolution (*vivartavāda*) with a view to show the process in which the world with its *modus operandi* is said to have evolved in the shape of meaning from the imperishable *Śabdatattva* of the essence of the *logos.*[107] Pandey presents Bhartṛhari's thought as one which adopted the doctrine of *avidyā* from Vedānta and is interpreted as *vivartavāda*. In his view, it is Bhartṛhari who introduced the idea of *vivarta* in the field of the Philosophy of Grammar. He finds it as the first philosophy of *vivarta.*[108] Murti compares the evolution of world in Bhartṛhari's system to that of the advaitic conception of *vivartavāda*. He says: "The Ultimate Ground is Transcendent, and generation of differences should be taken as a sort of emanation (*vivarta*) which does not affect the ultimate cause. Bhartṛhari sees the entire world as a non-transforming emanation of the non dual brahman, the WORD."[109]

Acknowledging the common authorship of the *Vākyapadīya* and its *Vṛtti,* Iyer asserts Brahman's powers as inexplicable. For him these powers are indescribable as identical or different from Brahman, as real or unreal, as compatible or mutually contradictory.[110] The very first verses refer to this relation of identity and difference between Brahman and its powers. Yet, as discussed elsewhere these powers are responsible for the appearance of differentiation in the undifferentiated Brahman. So there is provision in the *Vākyapadīya* for *avidyā* as the metaphysical principle of ignorance.

---

[106] K.A.S. IYER, *Bhartṛhari*, 135.

[107] P.C. CHAKRAVARTI, *Linguistic Speculations of the Hindus*, 28-29.

[108] S.L. PANDEY, *Pre-Śankara Advaita Philosophy,* 343-344.

[109] T.R.V. MURTI, *"Some Comments on the Philosophy of Language in the Indian Context,"* 331.

[110] K.A.S. IYER, *Bhartṛhari,* 134.

If so, can Bhartṛhari be designated as a *vivartavādin*? Bhartṛhari's argument that all effects partake in the nature of their cause and that the nature of the cause can be inferred by perceiving the element common to all its effects seems to be based on the *pariṇāmavāda*. But according to this theory the cause undergoes real change and appears as the effects which are real. Bhartṛhari on the other hand, does not accept a real change in Brahman which is changeless. He also does not seem to maintain that the effect namely the differentiated world is as real as Brahman.

On the contrary, Bhartṛhari seems to refer to the *satkāraṇavāda* or the theory according to which the cause possesses a higher degree of reality, for he defines the real as that which persists when all its forms come to an end.[111] Like gold which remains when its shapes disappear, Brahman, the ultimate source of all manifestations continues to exist when all its transformations come to an end.[112] The multiplicity being mere modification, is reducible to its cause and substratum, namely, Brahman which is real. However, *satkāraṇavāda* also implies differentiation between cause and effect which belongs to the realm of duality and is inapplicable to the non-dual and non-relational Reality.

Since change takes place only in entities conditioned by space and time, Brahman which is not affected by these categories is also beyond all change. From the transcendental point of view, Brahman cannot be a real or unreal cause, for it is ineffable. Brahman merely appears as the world of multiplicity under the operation of its powers. Real and unreal are not different entities but real, when not comprehended properly, is believed to be

---

[111]VP. III.2.11.

[112]VP. III.2.15.

unreal.[113] Brahman which is unchanging and undifferentiated is seen due to *avidyā* as the manifold world. Though the multiplicity cannot pertain to Brahman, it seems to belong to it.

The only concept that can explain the experience of differentiation in the non-dual and immutable ultimate Reality is that of unreal change. Knowledge and sciences which are based on differentiation are mere expedients for the ignorant and that by following such unreal path (*asatya vartman*) one reaches the Real.[114] In other words Real is in no way recipient of temporal limitations because its essence is absolutely undifferentiable. Taking up the question of change, the most fundamental form which makes the temporal distinctions possible, Bhartṛhari maintained that there is no change in the ultimate sense and that the true absolute transcends change. From the above discussions we come to the conclusion that, Bhartṛhari's position for the most part seems to be historically, logically and philosophically linked with *vivartavāda*, and is evident from the totality of his thought.

11. *ADVAITATTVA* OF *ŚABDATATTVA*

Bhartṛhari's philosophy has variously been called as the philosophy of Grammar, the philosophy of *Sphoṭa*, philosophy of Word and Meaning or the philosophy of *Śabdabrahman*. Since Bhartṛhari claims *Śabda* as the ultimate Reality of the world we prefer calling it *Śabdadvaita* or to use Bhartṛhari's own term *Śabdatattva vāda*. On the surface there are apparent parallels between the two schools of the Advaitin and the Grammarian. For both the Advaitin and the Grammarian the world of phenomenal differentiation appears as a free act of the Absolute, yet the Absolute is judged to be in no way diminished by the phenomenalization.

There is difficulty in understanding Bhartṛhari as influenced by

---

[113]VP. III.2.7.

[114]VP. II.238.

Advaitic thought, especially in his conception of *vivarta* as observed by some scholars. For, his relation to Advaita can scarcely be asserted if he lived and worked almost three hundred years before Śaṇkara. As Pandey observes the philosophy of Grammar has been interpreted as *vivartavāda* not under the influence of the Śaṇkara Vedānta. On the contrary it is the Vedānta of Śankara which adopted Bhartṛhari's doctrine of *avidyā*..[115] Dasgupta, explicitly speaks about *vivarta* as a concept which is considered to be of Śaṇkara, has its foundation already in the *Vākyapadīya* of Bhartṛhari.[116] Again Daniel Ingalls argues that Śaṇkara has adopted his *vivarta* doctrine from Bhartṛhari's thought.[117]

The *Śabdatattva* and its powers are all essentially identical and thus the Supreme Reality is, in the system of Bhartṛhari, one unqualified and undifferentiated unity of existence. Hence the position of the Grammarian is nothing but a non-dualism. It seems to me that Śaṇkara was aware of Bhartṛhari's thought and developed his own philosophy along the line with Bhartṛhari. However, Śankara, replaces Brahman in the place of Bhartṛhari's *Śabdatattva*, as the ultimate Reality. At the same time with the Mīmāṃsists, Śaṇkara also accepts the eternality of *śabda*. This would mean that there are two eternal entities- Brahman and *śabda*. But this is unacceptable when Brahman is defined as "absolute non-duality." This keeps the Advaitin definitely in confusion. But in Bhartṛhari's system he has cleverly avoided such a dilemma, through the identification of *śabda* and Brahman, thus propounding the doctrine of *Śabdabrahman*.

---

[115]S.L. PANDEY, *Pre-Śankara Advaita Philosophy*, 344.

[116]S. DASGUPTA, *Indian Idealism*, 195.

[117]D.H.H. INGALLS "The Study of Śaṃkarācārya," *Annals of the Bhandarkar Oriental Research Institute* 33 (1952), 1-14.; see also K.H. POTTER, *Encylopedia of Indian Philosophies*, Vol. III., 118.

12. CONCLUSION

Search for ultimate truth or ultimate Reality has remained the basic concern and the central problem for the Indian thinker throughout the history. Bhartṛhari asserts that whenever we assert of anything that it exists, its existence is of the nature of Brahman. Likewise, whenever we say of anything that it appears, the bare element of manifestation as abstracted from the content that is manifested is the Brahman. This shows, according to Bhartṛhari, that Being is as much immanent as transcendent with regard to ordinary experience: it is immanent as the indwelling condition of its possibility; transcendent inasmuch as in its purity it is beyond the limitations of contents encountered in experience. It would then seem that, the one Being is the universal substance; it is different from the finite substances *qua* substance in that the finite substances have qualities, whereas the one Being has no distinguishable quality; furthermore, of a finite substance, existence appears to be a predicate, whereas this one Being is the existence of which all finite substantial contents are, in reality, predicates.

Being is both immanent and transcendent, and such immanence-in-transcendence may have varying facets of identity and difference. Being itself is the unity and identity of what there is: in this very function, it may be taken as something over and above the multifarious entities; on the other hand it may appear as so deeply committed to the actual presence of this world that it cannot also represent its primeval ground and unity. Entities *are*, yet, Being alone *is*; Being transcends, and it has itself to be transcended. Bhartṛhari's position envisages one undifferentiated reality, the whole, paving way for the non-dualistic philosophy of Grammar. The undifferentiated reality is self-luminous and self-existent. It is the non-relational, non-dual reality. The empirical knowledge which is relational takes various relational forms according to the nature of the object, i.e., internal cognitions (subjective) and external sense perceptions (objective). It is through a synthetic analysis of these subjective and objective forms one reaches the non-dual Absolute which can be characterized as neither objective nor subjective. In other words, the ultimate reality is neither objective nor subjective. To reveal this truth, *Upaniṣads* purport that the individual (*Ātman*) is one with universal (Brahman), here and now. '*Ātman* is Brahman.'[118]

As an Advaitin Bhartṛhari shares this upaniṣadic view. It is evidenced when Bhartṛhari says that 'the ultimate substance is imperishable and is identical with what is called *ātman*.'[119] Hence he asserts that there exists an ultimate unity, deep-seated within all subjects and objects. The entire manifold universe ultimately and essentially expresses this unity. Both unity and diversity are terms relative to each other and have no individual independent existence of their own. Thus the Reality is non-dual having the nature of word as *Śabdatattva*.

Bhartṛhari occupies an independent position in the development of the Vedānta philosophy. For, he differs from the Mīmāṃsists who accept only the eternality of *śabda* and from the Advaitins who accept only Brahman as the eternal Reality. He takes an intermediary position by combining both these positions into one whole of *Śabdatattva*. Hence Bhartṛhari could be called both a pre-Śaṇkara linguistic philosopher of the Grammar School and an inter link between the philosophies of Mīmāṃsā and Advaita.

---

[118]Ma.Up. II.17

[119]VP. III.2.1.

CHAPTER TWO

ŚABDATATTVA: THE *SPHOṬA* OF LANGUAGE

1. INTRODUCTION

Among the various views that have been propounded by different schools on *śabda*, our present topic, the doctrine of *sphoṭa* upheld by the grammarian philosopher Bhartṛhari deserves special attention. The *sphoṭa* theory is a penetrative inquiry into the real nature of *śabda* (word) and *artha* (meaning). Although Bhartṛhari follows the tradition of Grammar School, he develops the *sphoṭa* doctrine probably against the background of the Mīmāṃsakā interpretation of *śabda* and *artha*. According to Mīmāṃsakās, since *Vedās* are eternal, the words of which they are composed also have to be eternal. In opposition to this theory of eternalness of the words Bhartṛhari holds that it is the meaning rather than the word that is eternal. All words in his view ultimately mean the Supreme Brahman which he calls *Śabdatattva*. In order to evince this truth he established the concept of *sphoṭa*, which looks for a more philosophical and transcendental grasp of the depth level of the vedic revelation.

Bhartṛhari presents his *sphoṭa* doctrine as a solution to the traditional problem of meaning. Hence in Bhartṛhari's view a word consisting of letters and syllables cannot directly convey its significance. A word or a sentence is to be considered not as a concatenation made up of different sound units arranged in a particular order, but mainly as a single meaningful symbol. According to him there is an eternal element called *sphoṭa* corresponding to each word which underlies the constituent letters and has no component parts. It is this unitary, unchanging *sphoṭa* latent in the word, which, being evoked as the letters and syllables are successively uttered or heard, produces the cognition of its meaning.

The term *sphoṭa* is derived from the root *sphuṭ* which means 'to burst' and it is defined in two ways. In its linguistic sense it is normally defined as 'that from which the meaning bursts forth', i.e., shines forth, in other words the word as expressing a meaning. Secondly it is defined as an entity which is manifested by the letters.[1] Nāgeśabhaṭṭa, a later Grammarian, in his *Sphoṭavāda* defines *sphoṭa* as: *'sphuṭati prakāśate 'rtho ' smād iti sphoṭaḥ'* as meaning bursts out of it or made evident from it, therefore it is called *sphoṭa*.[2] Coward traces the original conception of *sphoṭa* far back into the vedic period where *vāk* or speech was considered to be a manifestation of the all pervading Brahman, and the *Praṇava* (*Aum*) regarded as the primordial speech sound from which all forms of *vāk* were supposed to have evolved.[3]

Tradition ascribes the authorship of *sphoṭa* theory to sage Sphoṭayana. However, it is in the *Mahābhāṣya* of Patañjali that the first mention of term *sphoṭa* occurs. Patañjali uses the term *sphoṭa* as the permanent element in the word.[4] And in Bhartṛhari's view *sphoṭa* is an indivisible entity which is the real meaning bearing whole. *Sphoṭa* as a meaning principle works identically everywhere and in everybody's speech. Ganguli introduces the *sphoṭa* doctrine in full conjunction with the concept of *Śabdabrahman*. He says:

---

[1]K. RAJA, *Theories of Meaning*, 98.

[2]NAGEŚABHTṬA, *Sphoṭavāda* (Madras: Adyar Library 1956), 5.

[3]H.G. COWARD, *The Sphoṭa Theory of Language* (Delhi: Motilal Banarsidass 1980), 71-72. H.G. COWARD, *The Sphoṭa Theory of Language*.

[4]P.C. CHAKRAVARTTI, *The Philosophy of Sanskrit Grammar* (Calcutta: University of Calcutta 1930), 99.

> In its metaphysical aspect *Sphoṭa* is a hypostasized First Principle, a primordial entity which projects itself into the phenomenal world of cognition, meaning and language. The metaphysical Reality of Bhartṛhari is an integration of these three in one, and is conceived as the Final Substance which is the metempirical basis of the empirical world. ...Bhartṛhari... should call his First Principle *Śabda-Brahman* or the Primordial Word. It is called *Sphoṭa* because our epistemic universe gets its expression and emanation from it.[5]

Various scholars have attempted to define the term *sphoṭa*. Thus, according to G. Sastri, the original conception of *logos* best conveys the meaning of *sphoṭa*. He states: "...the act that *logos* stands for an idea as well as a word wonderfully approximates to the concept of *sphoṭa*."[6] Radhakrishnan presents *sphoṭa* as an indivisible idea with its dual form of *śabda*, word and *artha*, meaning.[7] T. Manickam sees *sphoṭa* as an inner perception which being a unitary principle links the subjective realms of cognition.[8] Along with Greek conception of *logos* Coward finds the Platonic notion of the innate idea also as aproximateble to *sphoṭa*.[9] Sometimes the word 'symbol' is used to denote the technical term *sphoṭa*. As K. Raja suggests the term symbol seems to be more ambiguous and *sphoṭa* should be understood more in the sense of a linguistic sign.[10] All these views point to one and the same *sphoṭa*, a latent unitary medium which forms the content of all words.

2. *SPHOṬA*: THE UNITARY MEDIUM OF EXPRESSION

Bhartṛhari begins with the treatment of *sphoṭa* which is the conveyor of real meaning from *Vākyapadīya* I.44. He says: "In

---

[5]H. GANGULI, *Philosophy of Logical Construction*, 60.

[6]G. SASTRI, *Philosophy of Word and Meaning*, 102-103.

[7]S. RADHAKRISHNAN, *The Principal Upaniṣads*, 674.

[8]T. MANICKAM, "Towards an Indian Hermeneutics of the Bible," *Jeevadhara* 17/68 (1982) 102.

[9]H.G. COWARD, "Language as Revelation," *Indian Philosophical Quarterly* 3 (1976) 450.

[10]K. RAJA, *Theories of Meaning*, 97.

the words which are expressive the learned discern two elements: one (*sphoṭa*) is the cause of the real word which, the other (*dhvani*), is used to convey the meaning."[11] Thus Bhartṛhari advocates that a word (*śabda)* has two aspects, namely, the word sound (*dhvani*) and word meaning (*artha*). Although they may appear to be essentially different, they are really identical. According to Iyer they are differentiated in the mind and yet integrated like two sides of the same coin which constitute the *sphoṭa.*[12] What Bhartṛhari emphasizes is the meaning bearing or revelatory function of this two sided unity, the *sphoṭa*, which he maintains is eternal and given in nature. The apparent difference is seen to result from the various external manifestations of the single internal *sphoṭa.*

The real expressive word, the indivisible *sphoṭa* is over and above the sounds which are many in number and are uttered by the speaker in a temporal sequence. This means that the separate letters of a word or words of the sentence merely manifest the *sphoṭa* or meaning-whole. As a rule, the sounds which are uttered by the speaker manifest the expressive word, which already exists in the hearer. In other words, when one utters a word which is in his mind, he produces a sequence of different sounds in order to make a sense out of many words. So it is not the manifesting sounds which convey the meaning but the indivisible sequenceless word that is hidden behind the visible words. That is why the listener though first hears a series of sounds he ultimately perceives the utterance as a unity.

As T. Patnaik points out Bhartṛhari assumes that the intention of the speaker to say something and the understanding of the utterance by the hearer is based on the common grasping of the

---

[11]*dvau upādānaśabdeṣu śabdau śabdavido viduḥ /*
eko nimittam śabdānām aparo'rthe prayujyate // VP. I.44.

[12]K.A.S. IYER, *Bhartṛhari*, 178.

content of speech.[13] Supporting the Grammarians' position, Madhva Acharya in his *Sarva-Darśana Saṃgraha*, says: "... as the letters cannot cause the cognition of meaning, there must be a *sphoṭa* by means of which arises the knowledge of the meaning, and this *sphoṭa* is an eternal sound, distinct from the letters and revealed by them, which causes the cognition of the meaning."[14] In his twofold analysis of language as uttered speech and as the bearer of meaning Bhartṛhari transgresses the level of uttered language to find its connection with thought and ultimately with the unifying ground of meaning. Here J. Brough's attempt to present *sphoṭa* as a theory of a language symbolism rather than a mystical entity, i.e., considering the word for linguistic purposes as a meaning-bearing unit, a single symbol whose parts are not relevant to its qua symbol is really appreciable.[15] Nevertheless, from the point of view of Bhartṛhari *sphoṭa* seem to represent both the linguistic symbol as well as the transcendental reality.

3. MANIFESTING PROCESS OF *SPHOṬA*

The whole word or idea exists in the mind of the speaker as a unity or *sphoṭa*. It is as it were an immediate intuition without any subject-object distinction. But when the speaker utters it, he produces a sequence of different word-sounds so that what is really one - the *sphoṭa* or whole idea - appears to be many. Thus in the process of manifestation each sound uttered by the speaker reveals *sphoṭa*, the first one vaguely, the second one clearly and so on until the last one helped by the impressions left by the

---

[13]T. PATNAIK, "Language, Thought and Communication," *Indian Philosophical Quarterly* 15/3 (1988) 358.

[14]MADHVA, *Sarva Darśana Saṃgrahagraha*, tr., E.B. COWELL & A.G. GOUGH, (New Delhi: Cosmo Publications 1976), 211.

[15]J. BROUGH "Theories of general linguistics in the Sanskrit Grammarians," in *A Reader on the Sanskrit Grammarians,* ed., J.F. STAAL (Delhi: Motilal Banarsidass 1985), 402-414.

previous cognitions reveals it in all its clarity and distinctness.

The cognition produced by each sound taken by itself is unnameable. Because it is not enough to identify *sphoṭa*. But each one of them is helpful in the rising of the final clear perception of it.[16] The previous cognition also had the *sphoṭa* as their object but the cognition of it was vague and that is why they had the form of the sounds. But when the final cognition reveals the *sphoṭa* in all its clarity and distinctness, it has no longer the form of the sounds. However, what Bhartṛhari means by early vague cognitions is not clear from the *Vākyapadīya*.

The process of speaking, as Coward observes, is reversed from the point of view of the listener. The listener, although at first hearing a series of uttered words, ultimately perceives the uttered sentence as a unity- the same unity or *sphoṭa* from which the speaker began- and it is then that the moment of communication between speaker and listener occurs.[17] When the *sphoṭa* in the form of sentence is being manifested by the sounds, there is, at first a cognition of parts in the form of phonemes and individual words. Nevertheless, the sounds which reveal the *sphoṭa* cannot constitute parts of it. They function as mere signs or diacritical marks which reveal the identity of the words.[18]

Another important aspect of *sphoṭa* is that though it is indivisible and timeless it can be revealed only by means of the sound in a time-series patterns. Concerning this issue Raja says that the time order is merely a means for revealing the timeless *sphoṭa* which he calls the *signifiant.*[19] *Vākyapadīya* says that we are capable of perceiving *sphoṭa* directly.[20] So we perceive *sphoṭa* when it comes within the range of worldly usage as

---

[16]VP. I.83.

[17]H.G. COWARD, "Language as Revelation," 451.

[18]K. RAJA, *Theories of Meaning*, 130.

[19]K. RAJA, *Theories of Meaning*, 124.

[20]VP. I.85.

differentiated word elements. Iyer sees *sphoṭa* very much like consciousness itself. He says:

> Consciousness is really one, but it becomes differentiated when it illuminates the different objects of the world. It seems to take on the form of the different objects which it reveals. It is only when it does so that it enters into worldly usage. Similarly, as long as the word is only within us, one and undifferentiated, it cannot enter into verbal usage... Thus consciousness and the word have to go through an unreal differentiation before they can play their part in the world. In fact, mind and the word are only two aspects which the *Śabdatattva* (word principle) within us assumes when it evolves.[21]

Though *sphoṭa* is a unitary meaning-whole, it may have as many different manifestations as there are speakers to utter it - each with his own peculiar way of putting it into words, his own speed of speaking, his own accent, etc. This means even when the external manifestations are different the inner meaning of the *sphoṭa* remains unchanged. K. Kadankavil points out that according to Bhartṛhari our words and sentences have to be the manifestations of a real language which is the true bearer of meaning. In other words the whole world of words and objects has to be a manifestation of an Absolute Word. Our speech imperfectly imitates it. Hence speech, in its proper sense, is one whole meaning without parts.[22] One needs to go through the unreal manifestation before he reaches the partless whole. However, the sages to whom the Eternal Light of *Śabdabrahman* has manifested itself, always understood *sphoṭa* without going through such an unreal differentiation of word. To them the first sound manifests *sphoṭa* fully and completely.[23]

---

[21]K.A.S. IYER, *The Vākyapadīya*: *Some Problems* (Poona: Bhadarkar Oriental Research Institute 1982), 32.

[22]K.T. KADANKAVIL, *The Quest of the Real* (Bangalore: Dharmaram Publications 1975), 145.

[23]G. SASTRI, *Dialectics of Sphoṭa*, 11.

4. *VĀKYA- SPHOṬA*: THE MEANING BEARING UNIT

The discussions in the second *kāṇḍa* of *Vākyapadīya* occupies a central question whether it is the sentence (*vākya*) or the word (*pada*) that constitute the primary unit of language. *Kārika* II.57 calls the upholders of the sentence *vākyavādins*, i.e., the Grammarians and the upholders of the word as *padadarśins*, i.e., the Mīmāṃsakās.[24] In contrast to the Mīmāṃsakās who conceive the sentence meaning or the complete thought as resulting from the summation of the individual meanings, Bhartṛhari understands *sphoṭa* primarily as an indivisible sentence which is expressive of sense.[25] Technically it is also called *akaṇḍa vākyasphoṭa*. Many of the Indian philosophers held the view that a sentence is the unit of meaningful expression, and that a sentence is a further (semantically) indivisible unit. The indivisible sentence theory (*akaṇḍa-vākya vāda*) is succinctly stated by Bhartṛhari in the well-known verse. "*Pade varnā na vidyante*

---

[24]For a contextualized study of the theme see J.E.M. HOUBEN, "Who are Bhartṛhari's Padadarśins? On the Development of Bhartṛhari's Philosophy of Language," in *Bhartṛhari: Philosopher and Grammarian*, eds., S. BHATE and J. BRONKHORST (Delhi: Motilal Banarsidass 1994), 155-169.

[25]The Indian philosophers carried on discussions advanced by arguments and counter-arguments over several centuries about the related designation (*anvitābhidhāna*) and the relation of designata (*abhihitānvaya*) theories with seemingly endless arguments. The theory of related designation (*anvitābhidhāna*) is the view that a word has meaning only in so far as it functions in a sentence in co-operation with the other words that belong to the sentence so that a word signifies its meaning only as related to the meanings of those other words. The theory of relation of the designata *(abhihitānvaya)* is the view that words convey their own meanings, and that the sentential meaning is composed of word-rneanings as related in a certain manner. According to Bhartṛhari who hold the sentence theory, the sentence is the basic, indivisible semantic unit. Word meaning is a mere theoretical construct of the science of grammar. Siderits calls it the problem of sentential unity. See, M. SIDERITS, *Indian Philosophy of Language* (Dordrecht: Kluwer Academic Publishers 1991), 9.

varneṣvavayavā iva vākyātpadānā matyantam praviveko na kascana" which may be rendered as 'the phonemes do not exist in a word, just as the phonemes themselves have no parts. The words do not have complete distinction from the sentence. A sentence is neither a unified collection nor an ordered series of words, nor is it sufficient to add the standard requirements (on many Indian theories) of contiguity, expectancy appropriateness. A word is an artificial construction, an isolated word a fiction. A sentence alone is the unit of utterance, a single indivisible entity with a single undivided meaning that is grasped as a unity in a flash of insight *pratibhā.*

In the first chapter of the third book of *Vākyapadīya*, Bhartṛhari restated his position as follows: Just as the *padavādins* (those who regard the word as the primary indivisible unit) consider word-constituents, such as roots and suffixes, to be mere fictitious abstractions from words, so also the *vākyavādins* (those who regard the sentence as the indivisible unit) consider words to be imaginary abstractions from the sentence.[26] Thus, in the beginning of the fourth chapter of the third book, Bhartṛhari rephrases his theory as follows: For the grammarians (i.e., for the *vākyavādins*) the word-meanings are fictitious constructions from the sentence-meanings (which are to be regarded as wholes), just as for the *padavādins*, the meanings of roots and suffixes are mere abstractions from the indivisible word-meanings.

The sentence-meaning, according to Bhartṛhari, is an indivisible *samvit* - one whole cognitive content - from which different word-meanings - also called ontological categories - are abstracted by imaginative construction.[27]Bhartṛhari has definitely stated that a word or letter has no reality by itself apart from that of a sentence. The basic logic behind his thinking is that the whole is prior to the parts. Iyer's observation adds to the point:

---

[26]VP. III.1.1.

[27]VP. III.4.1-2.

> The main thesis of the *Vākyapadīya* of Bhartṛhari is that the word is the *sphoṭa* of which the main form is the sentence which is indivisible, both in its sound aspect and in its meaning aspect. It is that which bears the meaning and which is, therefore expressive (*vācaka*). That alone is used for the purpose of communication (*vyavahāra)* and therefore, real.[28]

Bhartṛhari makes his stand clear by observing that man does not speak in individual words. For it is the whole idea or complete thought which is the expressive aspect of *śabda* (*sphoṭa*). He gives an example for it. Even when a word is used merely in the form of a substantive noun (e.g. 'tree'), the verb 'to be' is always understood so that what is indicated is really a complete thought (e.g. 'It is a tree').[29] He continues further saying that the real unit of language is the sentence, and that for pedagogical purposes words are abstracted from the sentence and ascribed a meaning. The *Vṛtti* says, this artificial division of the sentence and the sentence-meaning is a means to understanding of the indivisible sentence and the sentence meaning.[30] In short for Bhartṛhari the sentence as a whole has the meaning as a whole.

Completeness of meaning is then the test of a sentence. It points to the fact that even a single word can be a sentence if it achieves the completeness of meaning in some way. For the same reason, sentences which require some other word or words for the completion of its meaning is incomplete. It seems that Bhartṛhari really intended it. That may be the reason by he suggests that the meaning of sentences consisting of single words should be determined by purpose and context.[31] It is a plain truth that without the unifying function of the meaning of a sentence words cannot stand; they will scatter and wither away. This means the

---

[28]K.A.S. IYER, "Bhartṛhari on the Primary and Secondary Meanings of Words," *Indian Linguistics* 29 (1968) 47.

[29]VPV. I.24-26.

[30]VPV. II.13.

[31]VP. II.330.

meaning has to stand as a principle at the root of words and sentences.

Later Grammarians speak about various kinds of *sphoṭa*. Thus Bhaṭṭoji Dīkṣta in his *Śabdakaustubha* enumerates no less than eight varities of *sphoṭa*: *varṇa-sphoṭa, pada-sphoṭa, vākya-sphoṭa, varṇa-jātisphoṭa, pada-jātisphoṭa, vākya-jāti-sphoṭa, akhaṇḍa-apada -sphoṭa* and *akhaṇḍa-vākya-sphoṭa*.[32] It should be noted, however that of all these eightfold divisions it is the *vākyasphoṭa* which alone represents *sphoṭa*. Other divisions are there only for practical purposes. That is why Bhartṛhari in a spirit of accommodation admits the need for analyzing the sentence into words, letters etc. From the grammatical point of view Bhartṛhari admits the need for classifying sentences for it enables the beginners to comprehend the sentence as a totality by itself.

The difficulty with individual word according to Bhartṛhari is its indefiniteness. Words are apparently imperfect because they depend upon a higher unity of sentence. Thus Pandeya points to the fact that the *sphoṭa* owing to its power is manifested as words on the one hand and as objects on the other. Words and objects exist, but they exit not as they are appearing but as *Sphoṭa*. ...Words are apparently imperfect because they depend upon a higher unity of sentence.[33] He continues thus:

> He (Bhartṛhari) says that words that we speak are momentary and they cannot be the bearers of meaning. They are only manifestations of a real language which is the true bearer of meaning. This language is the highest reality and the whole world of words and objects is a manifestation of this Absolute Word. Our speech imperfectly imitates this.[34]

As has been seen elsewhere Bhartṛhari admits the fact that no

---

[32]BHAṬṬOJI DĪKSITA, *Śabdakaustubha* (Benares: Chowkhamba Sanskrit Series 1933), 7. See also NAGEŚABHṬṬA, *Sphoṭavada.*

[33]R.C. PANDEYA, *Problem of Meaning*, 265.

[34]R.C. PANDEYA, *Problem of Meaning*, 18-19.

knowledge is possible without words. Words as carriers of meaning are really related to meaning. He is of the opinion that there is a natural fitness (*yogyatā*) between the word and meaning. Hence the purpose of Bhartṛhari seems to be not to disparage the value of words but to deny the transcendental validity of parts as they are always parts and never wholes.

5. *SPHOṬA* AND *ARTHA*

As has been observed it is the indivisible sentence which when manifested through sounds conveys the meaning (*artha)*.[35] But how a sound is related to a thing or how the word uttered by one becomes so powerful and active as to convey the intended sense to the listener is a question yet to be answered. A satisfactory answer to this question cannot be given unless we can sufficiently prove the existence of some kind of relationship between the two. For Bhartṛhari the relation between word and meaning is that of a natural relationship. Since words come into use, they are related with their meanings. In *Vākyapadīya* III.3.29 Bhartṛhari defines this relation as the natural fitness (*yogyatā*) of the words. He assumes that the word is never devoid of its *yogyatā* which is a given relation between word and meaning and is not shifted.

Bhartṛhari considers it as the most plausible explanation that could appeal to one's reason. That is why he assumes with boldness and perspicuity that *śabda* and *artha* are nothing but different aspects of one and the same thing. He elucidates it thus:

> According to them, the inner meaning, (that is, the sentence-meaning) is manifested by parts of it. Word and Meaning (that is, Sentence and

---

[35] *Artha* means both the linguistic meaning as well as concrete object or really existing entity. In general, Bhartṛhari and the Grammarians tend to accept *artha* in the first place as linguistic meaning. While the other meaning cannot be excluded.

sentence-meaning) are inseparable (*apṛthaksthitau*) divisions of one Inner Principle.[36]

It points to the fact that *śabda* and *artha* both remain inseparable in the mind prior to their outward manifestation. And when it is manifested, its two aspects the word aspect and the meaning aspect appears to be differentiated. So what Bhartṛhari emphasizes again and again is the fact that a word has a double power; it can convey an idea of the form of an expression as well as its content. In other words, words have the power revealing their objects or meanings and that of revealing themselves. Bhartṛhari compares this double power of words to that of a light which reveals itself and at the same time reveals other things.[37] Hence these two powers of a word are not really different from the word itself.

In *Sambandhasamuddeśa* verse 38 Bhartṛhari denies that any person would be able to create a relation between words and meanings if these words do not already have a relation i.e., a relation with some meaning. In other words, Bhartṛhari denies the possibility of pure convention. Again verses 1- 4 and 29 make it clear that for Bhartṛhari the relation between *śabda* and *artha* is a relation that is given. In verses II.118 and III.3.20 Bhartṛhari discusses further the question of natural relation in a convincing manner.[38] As Iyer sees, the two, *śabda* and *artha* stand in relation

---

[36]*arthabhāgaiḥ tathā teṣām āntaro'rthaḥ prakāśate /*
ekasyaiva ātmano bhedau śabdārthau apṛthaksthitau // VP. II.31

[37]*grāhyatvam grāhakatvam ca dve śaktī tejeso yathā /*
tathaiva sarvaśabdānām ete pṛthag avasthite // VP. I.55.

[38]The relation between a word and its meaning stands as a central problem in the philosophy of language. Bhartṛhari makes an excellent philosophical reflection on the analysis of the problem of word-meaning relation in the *Śambandha Samuddeśa* section of the third *kāṇḍa* of the *Vākyapadīya*. For a detailed study, see D.N.TIWARI, "Bhartṛhari's Philosophy of relation between word and meaning." *Journal of Indian Council of Philosophical Research* 11/2 (1994) 43-54.

to each other, they face each other and it is then that they are said to be *vācaka* (the expressive word) and *vācya* (the expressed meaning).[39]

Bhartṛhari has also shown the mutual causal connection between *śabda* and *artha*. They are said to be cause (*kāraṇa*) and effect (*kārya*). Again they stand towards each other as illuminator (*prakāśaka*) and the illuminated (*prakāśyatvam*).[40] In so far as the word conveys the object, the meaning and its own form, it is like knowledge which reveals itself while revealing the object.[41] That being so, the word cannot reveal the object or convey the meaning unless it reveals itself first. To put it differently, nobody can understand a meaning from the word, unless he understands or recognizes the word first.[42] That is why in ordinary communication when the own form of the word is not understood one asks the speaker, 'what did you say'?[43]

In this way Bhartṛhari very clearly explains the inseparable relation of *sphoṭa* and *artha*. Thus, in such a relation Bhartṛhari has found a fundamental solution of why things are invariably signified by their corresponding words. This undivided character of *śabda* and *artha*, says Chakravarti, comes to one's comprehension only in a stage of higher spiritual cultivation when all forms of dualism disappear or merge into the unity of the Infinite.[44]

6. *SPHOṬA* AND THE EVOLUTION OF LANGUAGE

According to Bhartṛhari, there are three levels of languages through which *śabda* or *vāk* passes whenever one speaks. *Sphoṭa*

---

[39]K.A.S. IYER, *The Vākyapadīya: Some Problems*, 17.

[40]VP. II.32.

[41]VP. I.50.

[42]VP. I.56.

[43]VP. I.57.

[44]P.C. CHAKRAVATRI, *The Linguistic speculations of the Hindus*, 335.

which is at first quite internal, is gradually externalized for the purpose of communication. The following *kārikā* expresses the evolutionary stages of *śabda* as follows:

> This Science of Grammar is the Supreme and wonderful source of the knowledge of the threefold word, comprising many paths, of the *Vaikharī* (the Elaborated), the *Madhyamā* (the middle one) and the *Paśyantī* (the seeing One).[45]

These stages namely, the *vaikharī*, *madhyamā*, and *paśyantī* are not defined or described anywhere in the *kārikās*. It is the *Vṛtti* which gives some information about them. *Paśyantī*, the first stage is one and absolutely free from all differentiation and sequence. It is so subtle that it cannot be perceived by senses. It is indivisible, it is the inner light, it is the subtle word and it is imperishable. At this stage, no distinction is possible between sound and sense.[46] In a word it is the pre-mental stage which represents the ultimate Reality. *Vṛtti* on *Vākyapadīya* I.14 presents *paśyantī* as the supreme Reality, *Śabdabrahman*, which is identified with *pratibhā* the flash of insight of the principle of consciousness. As Sastri points out, there is a section of Grammarians in whose view, the ultimate Reality is represented by *para* and not *paśyantī* which in their opinion, is the first manifestation of the Eternal *Verbum* (*parāvāk*) while *madhyamā* and *vaikharī* are the two subsequent manifestations.[47] Of course, there is mention, even in the *Ṛgveda*, of four different kinds of *vāk*[48] and even the *Vākyapadīya Vṛtti* quotes it. It seems that some of the later Grammarians also upheld the fourfold division of speech, namely, *para*, *paśyantī*, *madhyamā* and *vaikharī*. Nevertheless, Bhartṛhari does not make use of the term *para*

---

[45] *vaikharyā madhyamāyāś ca praśyantyāś caitad bhūtam / anekatīrthabhedāyās trayyā vācaḥ param padam* // VP. I.143.

[46] VPV. I.143.

[47] G. SASTRI, *Philosophy of Word and Meaning*, 68.

[48] *Ṛgveda* I.164.45.

nowhere in the *Vākyapadīya*. It points to the fact that for Bhartṛhari it is *paśyantī* which represents the ultimate Reality.

The second stage *madhyamā* is purely mental and is not audible to others. Being one with the mind which is sequenceless, it is also sequenceless, but appears to have sequence. In this stage, the word and the meaning are differentiated from each other, but each one is still a unity.[49] It could be compared to the *vākya-sphoṭa* in its mental separation into sentence meaning and a sequence of manifesting word sounds, none of which have yet been uttered. In other words it is mental and conceptual in nature. It represents the manifestation of meaning-essence on the level of thought. All the elements linguistically relevant to the uttered speech are present in his stage in the latent form.

Vaikharī, the third stage, is the word that is audible to others. *Prāṇa* (breath) plays a great part in it and so it has definite sequence and form. Here both the word and the meaning are fully differentiated. At this stage all the individual speech peculiarities of the speaker are present in the utterance.[50] The *vaikharī* is, however, distinguished from the rest by being perfectly audible and capable of being expressed through the medium of letters. This is, therefore, known as the popular form of speech current in human society. It is the actual utterance, expressing the intention of the speaker and understood by the hearer.

Bhartṛhari is emphatically of the opinion that no knowledge whatsoever is possible without the operation of the word or *sphoṭa*.[51] He does not evidently make any distinction between knowledge and word. To him they are only different in form but not in essence.[52] He goes on to say without any reservation that knowledge is translated into word. Words that we speak come

---

[49]VPV. I.143.

[50]VPV. I.143.

[51]VP. I.123.

[52]VP. II.31.

out as an embodiment of thought, which then, is called language. So language grows with thought, and in the reverse thought also grows with language. In this process of mutual transformation, the internal knowledge, holds Bhartṛhari, assumes an audible form and all ideas are communicated to others.[53] It means that Bhartṛhari believed in the impossibility of thinking without speech. Murti puts it this way:

> It is not that we have a thought, well formed and complete, and then we seek a word to express it; or that we have a lonely word which we seek to connect with a thought. Word and thought develop together, or rather they are the expressions of one deep spiritual impulse to know and to communicate.[54]

Bhartṛhari posits that the communication takes place because the word itself contains an inner energy which seeks to burst forth (*sphuṭ*) into expression.[55] Coward observes two ways in which the energy of speech causes the phenomenalization of the *sphoṭa*. He says: "On the one hand, there is the pent-up potentiality for bursting forth residing in the *sphoṭa* itself, while on the other hand there is the desire of the speaker to communicate."[56]

7. CONCEPT OF *PRATIBHĀ*

Bhartṛhari's conception of *pratibhā* is one of the important issues to be discussed in this connection. The word *pratibhā* is derived from the root '*bhā*' which means to shine upon, to come in sight, flash upon the thought etc. Thus it has more than one signification in Indian literature. In Indian poetics, *pratibhā* is used to designate the extraordinary imaginative or innate powers which poets possess. *Pratibhā* in literary theory, according to Sreekantaiya is considered to designate fundamental element of

---

[53]VP. I.112-113.

[54]T.R.V. MURTI, "Some Comments on the Philosophy of Language in the Indian Context", 321.

[55]VP. I.51.

[56]H.G.COWARD, *The Sphoṭa Theory of Language*, 78.

creative poetic experience or the poetic imagination.[57]

In the absence of any exact equivalent in English that would precisely bring out the Indian conception of *pratibhā*, translators have used such words as intuition, imagination, inspiration, supersensual intelligence, etc. in accordance with Western Philosophy. As G. Kaviraj observes the *pratibhā* concept as a distinct faculty seems to have its origin in an attempt to explain phenomena which are beyond the range of the intellectual powers. As a flash of light, a revelation, it might be called the supersensuous and supra-rational apperception, grasping truth directly.[58]

Pratibhā of Bhartṛhari's conception comes closer to the term intuition. He calls it as an instantaneous flash of insight. Further, he compares it with the light of consciousness which is verbal in nature. This concept describes Bhartṛhari's insight into the nature of the relationship between word and thing. It is this *pratibhā* which helps one to overcome the dualistic mode of separation between word and thing. Hence he defines *pratibhā*, the flash of insight which is innate in every being thus:

> When the meanings (of the individual words) have been understood separately, a flash of understanding takes place which they call the meaning of the sentence, brought about by the meanings of the individual words.[59]

This means that though the words have different meanings when they are individually apprehended, but the moment they are united together in the form of a sentence we have a single sense which is distinct from the meanings of isolated words. And this kind of sense derived from the sentence as a whole is called *pratibhā*.

---

[57]T.N. SREEKANTAIYA, "Imagination in Indian Poetics," *The Indian Historical Quarterly* 13(1937) 59-84.

[58]G. KAVIRAJ, "The Doctrine of *Pratibhā* in Indian Philosophy", *Annals of Bhandarkar Oriental Research Institute* 5 (1923-1924) 1.

[59]VP. II.143.

E.Christe observes thus: "There is no interval between our hearing the words and understanding the meaning of the sentence, so there is no time in which to work this meaning out; it simply appears in the mind."[60] Hence *pratibhā* is the basic element in the comprehension of sense. What is aimed at through the immediate operation of *pratibhā* is the association of meanings which seem to have been otherwise unconnected.[61] It follows that the flash of understanding (*pratibhā*) is directly manifested by the use of words. Though the meanings of individual words are not real or rather have only a practical reality, they serve the purpose of bringing up *pratibhā.*

Sastri, quoting from Helārāja's commentary on the third section of *Vākyapadīya* says that *pratibhā* is of the nature of light (or consciousness) and joy and when it dawns upon us the luminous self shines in the mind as a novel revelation.[62] Helārāja, therefore, identifies *pratibhā* with *paśyantī.* The assumption seems to be true, for the basic meaning of *pratibhā* manifest something unexplainable. It is experienced by everyone within himself. Yet one fails to render an account of it.[63] Bhartṛhari also speaks of six different kinds of *pratibhā* as obtained by nature, by action, by practice, by meditation, by invisible causes and as handed down by the wise.[64] As discussed earlier *pratibhā* in its highest manifestation has been identified with *paśyantī* the divine intuition of the Absolute. But Bhartṛhari does not hesitate to equate it with the lowest intuition of birds etc.[65] Bhartṛhari claims that this intuitive knowledge is far more reliable than any other

---

[60]E. CHRISTIE, "Indian Philosophers on Poetic Immagination (*Pratibhā*)," *Journal of Indian Philosophy* 7 (1979) 173.

[61]VP. II.145.

[62]G. SASTRI, *Philosophy of Word and Meaning*, 9.

[63]VP. II.144.

[64]VP. II.152.

[65]VP. II.149-150.

kind of knowledge because it comes from within. It can arise in all sentient beings, for its root cause is the Word-Principle which is an integral part of sentience and hence present (potentially) in all such beings.

In fact the *pratibhā* experience is that of a unified psychic entity. And in this sense Bhartṛhari connects *pratibhā* with the *vākyasphoṭa*. So the mental perception of the *vākyasphoṭa* is a case of *pratibhā*. A sentence alone is the unit of utterance, a single indivisible entity with a single undivided meaning that is grasped as a unity in a flash of insight *pratibhā*. We have seen that although the meanings of the individual words are quite different from sentence meaning, it is through the meanings of the individual words that the *pratibhā* of the *vākyasphoṭa* occurs. Consequently *pratibhā* cannot be understood by appealing to the phenomenal facts of ordinary perceptions. So Coward says:

> More often, however, inference and reasoning may have to be applied to the words of the sentence so that the individual's cognition is brought to the level where the intuitive grasping (*pratibhā*) of the meaning whole (*vākya-sphoṭa*) becomes possible. It is in this sense that reason is taken as the precursor of the revelatory experience.[66]

It is the fundamental unity of the whole that shines forth as *pratibhā* at all levels. Hence *pratibhā*, as Carpenter suggests is not merely an insight into the meaning of a sentence but a more intense experience of the luminosity of Being itself.[67] It is the direct perception of the Word (*sphoṭa*) itself. Moreover, Bhartṛhari depends upon the assumption of previous existences in order to defend his conception of *pratibhā* logically.[68] Once this presupposition is accepted, then the conception of *pratibhā* becomes reasonable and logically consistent.

---

[66]H.G. COWARD, *The Sphoṭa Theory of Language*, 84.

[67]D. CARPENTER, *Revelation, History and the Dialogue of Religions* (New York: Orbis Books 1995), 74-75.

[68]VP. II.146.

8. *DHVANI* OF *SPHOṬA*

Along with the concept of *sphoṭa*, the meaning bearing principle in every language and *pratibhā*, the instantaneous insight of knowledge, Bhartṛhari's philosophy occupies another important term, namely, *'dhvani'* (sound). The *sphoṭa*, although it is unitary and eternal in nature, requires many imperfectly suggestive expressions before its meaning is fully grasped or intuitively realized in one's mind. These suggestive expressions, i.e., the uttered expression or the manifesting sounds produced by the articulatory organs are called *dhvani* in Bhartṛhari's language.[69]

Ordinarily the word *dhvani* is used to sense suggestion, implication etc. But for Bhartṛhari it represents the uttered sound units. In simple terms *dhvanis* are the individual letter sounds which suggest or point out something. This something which is indicated by the *dhvani* according to Bhartṛhari is the *śabda* (*sphoṭa*, the real word). So *dhvani* could very well be called as the external aspect of internal *sphoṭa*. It is the *kārikā* 47 of the first *kāṇḍa* of *Vākyapadīya* that directly provides the basis for interpreting *dhvani* in terms of suggestion. There Bhartṛhari says it is the spoken words that suggest the inner meaning (*artha*) and eventually evoke the unified *sphoṭa*.[70] Bhartṛhari further

---

[69]Ghoshal Sastri traces two parallel traditions of *Tantra śāstra* (Treatises of Indian Esoterism) and the *Vyākaraṇa śāstra* (Treatises of Grammar), out of which the concept of *dhvani* has been evolved. See G.N. SASTRI, *Elements of Indian Aesthetics*, Vol. I (Varanasi: Chawkhambha Orientalia 1978), 148. However, it is the Grammar School as Pandey asserts which is credited with expounding the doctrine of *dhvani* for the first time. See K.C. PANDEY, *Comparative Aesthetics*, Vol. I. (Varanasi: Chowkhamba Sanskrit Series 1959), 281.

[70]The Prabhākra school of Mīmāṃsakās consider *dhvani* to be included in the primary function *abhidha* (the primary meaning) itself; for according to them the meaning of a word is what is conveyed by it. Because there is no restriction to the scope of the significative force of a word. Some of the

distinguishes two kind of *dhvanis*, namely the *prakṛta dhvani* and the *vaikṛta dhvani*. He also records various views on *dhvani* other than that of his own. Thus in Bhartṛhari one comes across a variety of views on *dhvani*.

Now for Bhartṛhari *dhvani* and *sphoṭa* are the special names of the two aspects of the expressive word mentioned by him in *Vākyapadīya* I.44. Hence, *dhvani* is the name or the sound which the speaker utters and which are the manifestors of real word, namely *sphoṭa*. *Sphoṭa*, when manifested by the *dhvanis* conveys the meaning. So the idea that the real word is the *sphoṭa*, an indivisible entity over and above the sounds (*dhvanis*) which are uttered by the speaker and heard by the listener and which conveys the meaning according to convention is the basis of Bhartṛhari's linguistic outlook.

Dhvanis which manifest the *sphoṭa* have temporal distinction because they can be short, long or prolated depending on the speaker. Although the individual letter sounds (*dhvanis*) vary with the individual speaker, this does not matter since they aim at the manifestation of one changeless *sphoṭa*. Meaning in Bhartṛhari's view, therefore is not conveyed from the speaker to the hearer, rather the spoken words serve as stimulus to reveal the meaning which is potentially present within the consciousness of every hearer.[71]

9. *DHVANI - ŚABDA* RELATIONSHIP

The relation between *dhvani* and *śabda* is not the same as the relation between a cognition and its object. For, the form of a

---

Alamkārikas (Rhrtoritions) like Mukulabhaṭṭa include the suggestive power (*vyañjanā*) under *Lakṣaṇā* (secondary meaning) itself. In Bhartṛhari's view it is meaningless to discuss the primary and secondary meaning of an individual word; the sentence has to be considered as a whole, and in the particular context in which it is uttered.

[71] VP. I.85-86.

cognition and that of its object are perceived as distinct from each other. But *dhvani*, and *śabda* are perceived as one.[72] There is of course difference in opinion with regard to the *sphoṭa-dhvani* relationship. Bhartṛhari himself explains it thus: "According to some, the sound is cognized as identical with the word (*sphoṭa*); others think that the sound is not perceived at all, which others still think it is perceived as a separate entity."[73]

Vṛtti discusses the three different views mentioned in the above *kārikā*. According to the first view *dhvanis* are perceived as identical with *sphoṭa*. And it is the *sphoṭa* which is perceived as one with the *dhavanis*, so that the properties of the latter are wrongly attributed to the former. In the second view *dhvanis* act as the cause of the manifestation of the *sphoṭa*, without themselves being perceived as in the case of senses. The upholders of the third view consider the perception of *sphoṭa* and the *dhvanis* as quite distinct.[74] According to Pandeya, the articulate speech of Bhartṛhari's view only illuminates the *sphoṭa*, as a flash of light illuminates a jar in a dark room. The light does not produce the jar, the jar was there and the light only illuminated it.[75] That is why he considers the word expression (*dhvani*) as a relative term, meaning the expression of something.[76] Murti compares the *sphoṭa-dhvani* relation to that of the relationship between the soul and body. He says: "The word, like the soul, has a physical embodiment in the sound and is made manifest through the latter, but the conveyance of

---

[72]K.A.S.IYER, *Bhartṛhari*, 173.

[73]*sphoṭarūpāvibhāgena dhvaner grahaṇam iṣyate /*
kaiścid dhvanir asamvedyaḥ svatantro 'nyaiḥ prakalpitaḥ // VP. I.81.

[74]VPV. I.81.

[75]R.C. PANDEYA, *Problem of Meaning*, 264.

[76]R.C. PANDEYA, *Problem of Meaning*, 38.

meaning is the function of the word; the sound only invokes the word."[77]

In the *Vṛtti* on *Vākyapadīya* I.76, Bhartṛhari makes an inner distinction within the *dhvanis*. *Dhvanis* in their first moment, that is at the time of the manifestation of *sphoṭa*, are called *prākṛtadhvanis* (primary sounds). Those are called *prākṛtadhvanis* without which the form of the *sphoṭa* would remain unmanifested and therefore unperceived. As for Bhartṛhari the primary sounds can be short, long or prolated. Duration seems to be the basis of this distinction. These properties of the primary sounds are wrongly attributed to the *sphoṭa*. Pandey finds the spoken language and other means of expressions as shadows of the real mental language. Therefore says he:

> Our spoken language, which is regarded by the people as the only language, is only shadow language having no existence of its own. ... It is affective and communicative only because it is an expression of the real language. People mistake our spoken language to be the real language because they fail to recognize the reality behind it. The verbal form of language is not real language. It is only an expression (*dhvani*) of the language *par excellence.*[78]

As soon as we hear the primary sounds, we perceive the *sphoṭa*. But the perception of the *sphoṭa* does not disappear at once. It lasts a little while more in quick speech, a little longer in speech of medium speed and longer still in slow speech. As Iyer sees, according to Bhartṛhari the sounds which follow the succeeding moments are like echoes or reverberations (*anuraṇanārupāḥ*) of those of the first moment. They are the *vaikṛta dhvanis* (secondary sounds) and are the cause of the repeated cognitions of the *sphoṭa.*[79] The secondary sounds do not in any way affect the quality of the *sphoṭa* already manifested by the primary

---

[77] T.R.V. MURTI, "Some Comments on the Philosophy of Language in the Indian Context," 325-326.

[78] R.C. PANDEYA, *Problem of Meaning*, 80-81.

[79] K.A.S. IYER, *Bhartṛhari*, 174.

sounds. Hence they are external to the *sphoṭa*. In *kārikā* I.77 of the *Vākyapadīya* Bhartṛhari says thus: "After the manifestation of the *sphoṭa*, the secondary sounds cause difference in speed of utterance, but the essence of the *sphoṭa* is not affected by them."

It has also been said that each secondary sound among them has its series of effects lasting as long as the vibration lasts and helps in the manifestation of the *sphoṭa*. Just as from the flames of burning fuel proceed continuously other flames produced by them and manifest objects by their light, so is the continuity of sounds.[80] The secondary sounds therefore, are the cause of the repeated perception of the same *sphoṭa*. Thus, according to Iyer a slow utterance of 'a' means the perception of a series of a=a, a, a. The fast utterance of 'ā' means the perception of a series of 'ā'= ā, ā, ā.[81] K. Raja considers the personal variations in the mode of utterance such as those in speech, pitch and intonation, etc., as analogous to the *vaikṛta dhvani* of Bhartṛhari.[82]

In Bhartṛhari's view it is not only the *prākṛta dhvanis* or the normal sound - patterns which reveal the linguistic sign (*sphoṭa*), but even the *vaikṛta dhvanis* or the individual modifications of the sound may also have an important role in speech activity. Again he confirms the fact that even though the word is eternal it is perceived only when it is manifested by *dhvanis* which are not eternal entities, but our own creation and, therefore, admit of distinction based on time. These sounds are susceptible to increase and decrease, that is greater and lesser duration. As the primary sounds are susceptible to increase and decrease, they can be short or long, or prolated. The perception of the primary sounds, with their particular duration is not a process external to the perception of the word or *sphoṭa*, because the latter cannot be

---

[80]VPV. I.106.

[81]K.A.S. IYER, "Bhartṛhari on *Dhvani*," *Annals of Bhadarkar Oriental Research Institute* 46 (1965) 53.

[82]K. RAJA, *Theories of Meaning*, 313.

perceived apart from the former. Therefore any distinction based on time which is found in the former is transferred to the latter and becomes effective. This also explains how *śabda* and time are inter-connected.

Bedsides the *prākṛta-vaikṛta* distinctions Bhartṛhari also speaks of another distinction, i.e., the distinction between *dhvani* and *nāda*. Even adjectives *prākṛta* and *vaikṛta* are found added to *nāda*, as they are to *dhvani* without any apparent difference in meaning.[83] In the *Vṛtti* on *Vākyapadīya* I.47, *nāda* is looked upon as a gross form or an accumulation of *dhvanis*. As the *Vṛtti* observes, *sphoṭa* is manifested by the resonance (*nāda*) produced by the sounds caused by contacts and separations.[84] In the final analysis one finds that Bhartṛhari's conception of *sphoṭa*, *prākṛta dhvani* and *vaikṛta dhvani* are corresponding to the three fold nature of the revelation of speech, namely, *paśyantī*, madhyamā and the *vaikharī* stages.[85]

10. THE *DHVANI* OF AESTHETICS

The notion of *dhvani* expounded in the Grammar Philosophy, particularly by Bhartṛhari, seems to have been crystallized and assumed in as a central notion in Indian Aesthetics. It is in the *Dhvanyāloka* of Ānandavardhana (850 A.D.) we have a detailed analysis of the theory of *dhvani*. He picks up the notion of *dhvani* as the suggestive power (*vyañjanā*) of the spoken words, but also uses *dhvani* to refer to the thing suggested, namely, the principle poetic mood or meaning.[86] Raja presents Ananda's conception of *dhvani* thus:

---

[83]VPV. I.49.

[84]VPV. I.102.

[85]S.K. NANDI, *Studies in Modern Indian Aesthetics* (Simla: Indian Institute of Advanced Study 1975), 195-196.

[86]ĀNANDAVARDHANA, *Dhvanyāloka*, ed., P. SASTRI (Benares: Kashi Sanskrit Series 1940).

> ... just as the sounds of utterances (*dhvani* in the grammarian's sense) reveal the integral linguistic sign (*sphoṭa*), so also a good poem with its sound, as well as the literal sense, reveals, over and above the literal sense, a charming sense which has great aesthetic value.[87]

This suggested sense over and above the literal meaning is the most important element in poetry. So, *Dhvanyāloka* I.1 calls it as the soul of poetry (*kāvyāsyatmā dhvaniḥ*).[88] Sharma finds certain similarities in both Bhartṛhari and Ananda. For both the Grammarian and Ananda the suggested element, i.e., the *sphoṭa* and aesthetic emotion are *nitya* (eternal). So suggestion is not restricted to *anitya* (temporal) objects alone. Again they both perceive an order of realization of the suggester and the suggested. Another important aspect is that the suggester reveals itself along with the suggested, as in the case of a lamp which reveals itself while suggesting itself.[89]

A. Anand's attempt to confine *sphoṭa* concept to the level of meaning and his understanding of Ānandavardhana's *dhvani* as a step further from meaning to meaningfulness, effecting the experience of significance[90] seem to be really limited. Because it would be unfair to treat *sphoṭa* as something devoid of the emotional aspects of meaning. Although Bhartṛhari has not explored explicitly the emotional level of meaning in his *sphoṭa* theory, he has definitely intended such a meaning situation which virtually includes the idea as well as the emotion. And it is very well be understood from his conception of *pratibhā* which in its

---

[87]K. RAJA, *Theories of Meaning*, 283.

[88]*Dhvanyāloka*, 9. For an English rendering of the text see *Dhvanyāloka or Theory of Suggestion in Poetry*, tr., K. KRISHNAMOORTHY (Poona: Oriental Book Agency 1955), 1.

[89]M.M. SHARMA, *The Dhvani Theory in Sanskrit Poetics* (Varanasi: Chowkhamba Publications 1968), 37-38.

[90]A. ANAND, *Philosophical Implications of Dhvani* (Vienna: Gerold & Co 1984), 27-28.

final analysis is nothing but an aesthetic and emotional experience of meaningfulness.

According to Abhinavagupta (1014) another important literary critic the transcendental function of suggestion in the poetic experience removes the primordial veil of ignorance from our minds and thereby allows the bliss associated with the discovery of true meaning to be experienced. Such bliss is technically termed *rasa* (aesthetic emotion) and is held to be fully experienced when *dhvani* removes the veil of ignorance so completely that the self of the experiencer stands clearly revealed. However we do not have any discussion on these aspects of *dhvani* in the *Vākyapadīya*. Bhartṛhari restricted its scope to the suggestive function of sounds which also acts as a medium for the manifestative meaning.

11. CRITIQUE OF *SPHOṬA*

The view point of the Grammarian, which we have discussed so far is not without any difficulty. The Mīmāṃsaka and Nyāya Schools vehemently criticize the Grammarians view, i.e., the *sphoṭa* theory on vital grounds. Jayanta Bhaṭṭa of the Nyāya School in his work *Nyāyamañjarī*, opens his criticism by saying that the Absolute can never be the Word Principle which is the very thesis of the Grammarians. He declares, further that the Grammarians' conception of *Śabdabrahman* as the ultimate Principle is hardly valid and should not deserve any serious attention.[91] We need not propose to discuss the Nyāya criticism further, instead we take up the Mīmāṃsā criticism to some extent for the obvious reason that it is in opposition to the Mīmāṃsakās that the Grammarian developed his doctrine.

The *Śabarabhāṣya* on the *Mīmāṃsa Sūtra* I.1.5 makes it

---

[91] *Nyāyamañjarī* (Summary), tr., J. BHAṬṬACHARYA *et al.*, *Encyclopedia of Indian Philosophies* Vol. II, *The Tradition of Nyāya-Vaiśeṣika up to Gangeśa*, ed., K.H. POTTER (Delhi: Motilal Banarsidas 1978), 384-385.

sufficiently clear that the theory of *sphoṭa* has been discarded by the Mīmāṃsakās on the ground that it strikes at the root of some of the fundamental principles of the system of their philosophy. However, the chief opposition came, at a later time, from the Mīmāṃsaka, Kumārila Bhaṭṭa. Kumārila, in his *Ślokavārttika* dedicates an entire section to refute the *sphoṭa* theory.[92]

Kumārila points out that when each one of the several sounds manifests one identical *sphoṭa*, there is little justification for producing the subsequent sounds. The Grammarians meet the objection by saying that *sphoṭa* will be revealed on the very moment when the first letter is perceived; but, it becomes more and more definitely manifested with the perception of each one of the succeeding letters. And, so, Kumārila's apprehension that succeeding letters will be redundant, is absolutely without any foundation.[93] Sastri postulates that *sphoṭa* is no doubt presented to the mind even by the first sound. But the mind is so constituted that it must be trained to receive the full glimpse of *sphoṭa*.

In reply to this explanation, Kumārila says that what is manifested by each of the sounds is nothing but a letter. For instance in the case of the word 'cow,' the first sound does not reveal in an indistinct way the required *sphoṭa* in question nor the succeeding sounds reveal the same in a progressively distinct manner. For the three sounds reveal the three letters, 'c', 'o' and 'w' respectively. So, the opponent argues that the supposition of the initial appearance of *sphoṭa* as letters and the invariable subsequent manifestation of its reality as *sphoṭa* are mere

---

[92]KUMĀRILA BHAṬṬA, *Ślokavārtika,* tr., G. JHA (Delhi: Sri Satguru Publications 1983), 261-281. See also, G. SASTRI, "The Doctrine of *Śabdabrahman* -A Criticism by Jayantabhaṭṭa," *Indian Historical Quarterly* 15/3 (1939) 441-453.

[93]In analyzing the Mīmāṃsā criticism of the *sphoṭa* I have followed, to some extent the line of thought as presented in Sastri's excellent summary of the *sphoṭa* doctrine, in *A Study in the Dialectics of Sphoṭa,* 6-12.

postulations which lack proof. In the present case, however, we cannot satisfactorily explain why a particular *sphoṭa* will not be at first manifested as such but as something different from it. However, Bhartṛhari's theory has the sanction of experience. For it is never felt by us whenever a word is heard there is the recollection of letters into words as the Mīmāṃsakās say instead we perceive the word and its meaning as a whole. And Bhartṛhari calls this whole the *sphoṭa*.

Again, we cannot maintain that a particular object should at times invariably appear as something else. For example, a rope is not always mistaken for a snake. Bhartṛhari replies that it is a fact that objects do at times appear as something quite different in nature from themselves. It is a piece of common knowledge that a tree observed from a distance appears to be an Elephant. But on more close and careful observations at subsequent moments of time, the tree is recognized as such.

Finally, the opponent asserts that the manifestation of *sphoṭa* is untenable. Because the Grammarians maintain the relation between sounds and *sphoṭa* as one of the manifestor and manifested, it follows that they reside in the same locus. But, we know that *sphoṭa* lies in the mind while sounds are produced in the cavity of the mouth. It means that the aforesaid relation between sounds and *sphoṭa* is impossible. Bhartṛhari clarifies the point saying that it is only in the case of concrete substances (*mūrta*) that the imposed condition is necessary. But, as neither *sphoṭa* nor sound has an concrete existence, the said condition cannot be legitimately demanded.

The Grammarians, after the analysis of the question of word and meaning, firmly establish that it is neither the word nor the letter but the indivisible sentence (*vākyasphoṭa*) which is endowed with the power of expressing a coherent sense. As against the criticism of *sphoṭa* by other schools, Murti supports the Grammarian's view thus:

In linguistic apprehension, as in other cognitions, there is the

> interplay of two factors of different levels - the empirical manifold of sense-data (the separate letters of words in this case) and the transcendental of *a priori* synthesis of the manifold which alone imparts a unity to those elements which would otherwise have remained a mere manifold. I have interpreted the doctrine of *Sphoṭa* in terms of Kant's doctrine: that all knowledge is a synthesis of the manifold of sense-data, because the *Sphoṭa* functions exactly like a transcendental category of the Whole or Unity.[94]

The Grammarians attitude toward word is two fold, i.e., transcendental and empirical. From the transcendental point of view the ultimate Reality is the absolute word (*Śabdatattva*) which is devoid of all distinctions. And from the empirical point of view this *Śabdatattva* is subject to further and further divisions. But the word remains unaffected at all times. In fact it is the sounds which overshadow the real identity of word.

In contrast to the Mīmāṃsist empirical outlook of meaning the Grammarian's theory is more of a transcendental nature. In that sense, *sphoṭa* of the Grammarians may be described as the transcendental ground in which the spoken syllables and conveyed meaning find themselves united as word or *śabda*. By assuming *sphoṭa* as an indivisible entity, the Grammarians could solve all the difficulties with regard to the problem of sequence which was a staggering blow to Mīmāṃsist conception of meaning. In this respect Bhartṛhari's theory stands atop and is able to compete with any other linguistic theory.

12. *ŚABDATATTVA*: THE PRINCIPLE OF INTEGRATION

After having dealt with Brahman as the Word Principle, the *sphoṭa* of Language it remains to consider how Bhartṛhari sees the *Śabdatattva* as a principle of integration in our every day existence. Bhartṛhari describes Grammar as a doorway to liberation (*mukti*). He emphasizes in unambiguous language that

[94] T.R.V. MURTI, "Some Comments on the Philosophy of Language in the Indian Context," 329.

the ultimate purpose of the study of Grammar is to secure release from the bondage of ignorance. This release is to be reached by means of a communion known as *śabdayoga*, *śabdapūrvayoga* or *vāgyoga*.

Terms such as *mokṣa, nihśreyasa, apavarga* are used interchangeably by Bhartṛhari in the opening *kārikās* of the *Vākyapadīya* meaning liberation in its various nuances. Towards the end of the first *kāṇḍa* Bhartṛhari comes back to this idea and tells that the purification of the word (*śabdasaṃskāra*) is the means to the attainment of the Supreme Self who is the Word-Principle (*Śabdatattva*). In the vision of Bhartṛhari the very ontological reality of *vāk* in its various levels amounts to a description of the path by which *mokṣa* (ultimate union or realization which man can have with the *Śabdatattva*) may be attained. *Kārikā* I.14 of *Vākyapadīya* states: "It (Grammar)is the door to liberation, the remedy for all the impurities of speech, the purifier of all the sciences and it shines in every branch of knowledge."

The *Vṛtti* on the above verse makes it clear that the use of correct word not only reveals all knowledge, but also it simultaneously offers itself as a means towards acquiring the spiritual merit necessary for man to be united with *Śabdatattva*. It is at the level of *pratibhā* (intuition) that such union productive of *mokṣa* is depicted. Bhartṛhari advocates Grammar as the first step toward ultimate integration which consists in liberation.[95] By mastering the science of Grammar, one attains the Supreme Brahman.[96] This is to be ascertained through meditation known as *śabdapūrva yoga*.[97] In its literal sense the term *śabdapūrva yoga* would mean the meditation on the unity of the ultimate Reality

---

[95]VP. I.16.
[96]VP. I.22.
[97]VP.I.20.

which is the Word. Practice of this *yoga* helps one to realize that *vāk* or word is the foundational Reality and that the sounds, syllables and their succession are all just reflections of the self-same Reality. G. Sastri quotes from Puṇyarāja's commentary, wherein he describes the process of meditation and its culmination in the vision of the truth and final union with it as follows:

> The aspirant reaches the essence of speech the pure *verbum*, which lies beyond the vital plane, by withdrawing his mind from external objects and fixing it upon his internal nature. This entails the dissolution of temporal sequence of thought activity. The purification of the *verbum* (i.e., the eternal light of consciousness which ever shines within the subject) results from this and the aspirant enters into it after having served all his ties with the material objective plane. This leads him to the attainment of the internal light and freed from all bounds and limitations he becomes identical with the Supreme Light- the Eternal Word Principle - the undying and undecaying Spirit called *Śabdabrahman* or the Word-Absolute.[98]

Vāgyoga, therefore, is a kind of meditation which aims at raising the level of the consciousness of words to the highest stage of the Word-Principle. The complete absence of all sorts of differentiation and sequence is the chief characteristic of it. The *vāgyoga* demands a kind of *śabdasaṃskāra,*[99] i.e., the knowledge of purification of words. This purification consists in discarding of unchaste or corrupt (*apabhraṃśa*) words and adherence to chaste or correct (*sādhu śabda*) words which culminate in the attainment of the ultimate Reality.

In Bhartṛhari's view the correct use of words leads to spiritual merit where as the corrupt use does not.[100] So far as the conveying of meaning is considered he agrees that the corrupt forms are capable in conveying meaning even to one who does

---

[98]G. SASTRI, *Philosophy of the Word and Meaning*, 17.

[99]VP. I.132.

[100]VP. I.27.

not know grammar. These corrupt words are called *apabhraṃśa*. The word *apabhraṃśa* literally means distortion or deviation from the original. As G. Sastri observes, when syllables are distinctly pronounced and there is no distortion of any kind, it is called chaste or correct word (*sādhu śabda*); and when syllables are not distinct but confused it is called *apabhraṃśa* .[101] According to the *Vṛtti*:

> Once the essence of the word is purified by the form the correctness of which is established and particular merit is manifested by the disappearance of obstacles in the shape of incorrect forms, well-being (*abhyudaya*) is certain.[102]

As has been stated earlier, the essence of Bhartṛhari's view lies in the conception that the Absolute Existence itself manifests in the form of words and their meaning. There is no difference between them; Brahman Himself is the Word. Our speech imperfectly imitates this Absolute Word. Liberation consists ultimately in the understanding of this truth. When one fails to recognize this fact, he identifies real with wrong associations and relations: wrong associations are due to the state of ignorance (*avidyā*). Ignorance is just that and nothing more in Bhartṛhari's vision. When one is able to distinguish between what is right and what is wrong, he liberates himself from wrong that is unreal, and he would realize right which is real. So liberation consists in the distinguishing power which is arrived at through *śabdayoga*. *Śabdayoga*, is therefore, adherence to the truth of the *Śabdatattva*.

It is only the proper Grammar and proper use of Language that will lead us beyond the mire of confusion and wrong associations. One who has attained such a state of existence is a perfect man, who distinguishes truth, speaks truth and as a result acts truthfully. Perfection in thinking, speaking and acting makes one's life more and more integrated. Thus, *Śabdatattva* of Bhartṛhari's

---

[101]G. SASTRI, *Dialectics of Sphoṭa*, 83-84.

[102]VPV. I.131.

conception, in its final analysis acts like a principle of integration which culminates in the ultimate liberation. It is not a fancy theory and certainly not a discussion on the Absolute as completely out there as a pure transcendent beyond our reach. In this perspective Bhartṛhari was pretty much a humanist rather than a mystic.

Bhartṛhari's linguistic analysis is not a language game, but as it is basically linked with the spiritual discipline called *vāgyoga* it involves a theory of human life and a vision of life's end. This vision of life's end is called *vāgyoga*. *Vṛtti* on *Vākyapadīya* I.130 declares the *vāgyoga* as a process that involves the breaking of one's ego-sense, the sense of the 'I' and 'mine' and thereby realizing the non-differentiated Word which is the Absolute. *Vāgyoga* - the *yoga* of speech thus lends man to the complete comprehension of truth. The word which is within the self of the speaker is said to be a great "bull" (*vṛaṣbha*) with whom union (*sāyūjya*) is desired.[103] This is called *mokṣa* or the attainment of Brahman. It is interesting to note in this context Potter's observation that even the great philosopher Śaṇkara finds the mechanism of liberation ultimately as an act that requires speech.[104]

The question whether the supreme Reality of Bhartṛhari's system is of the nature of bliss, as that of the Vedāntins remains yet to be answered. The *Vākyapadīya* does not lend any positive support to the suggestion that Bhartṛhari identifies his Absolute with bliss. In Sastri's view Helārāja's identification of Eternal *Verbum* with *pratibhā* appears to suggest that the supreme Reality of the conception of the Grammarian is of the nature of pure bliss. Puṇyarāja, however, considers the Absolute not as

---

[103]VP. I.130.

[104]K.H. POTTER, *Encyclopedia of Indian Philosophies*, Vol. III. 54. For an excellent study on the Advaita perspective of Language see, J. GRIMES, *An Advaita Vedanta Perspective on Language* (Delhi: Sri Satguru Publications 1991).

bliss but as being and as consciousness, which is the chief good in man's life.[105] Thus through the advancement of *śabdapūrvayoga* Bhartṛhari has joined to that group of philosophers, for whom goal of philosophy is not mere intellectual satisfaction, but rather spiritual emulation.

13. CONCLUSION

Real is the essence of words (*Śabdatattva*) and is the Word (*sphoṭa*) itself. The Word is therefore called as that from which objects emerge: it is *sphoṭa*. *Sphoṭa* is one and does not involve any order like priority and posteriority. The unique concept of *sphoṭa* is an unparalleled contribution of Indian Mind. It may resemble the *apūrva* concept of the Mīmāṃsakas which acts as a link between what is heard and what is comprehended. Its identification with Brahman is the point where Grammar and Philosophy meet together. Its appearance is real because of the Real. It is like a tree and its branches. As the branches have no existence apart from the tree so also the appearance has no existence apart from the real. Ultimately words and objects are nothing but existence itself.

Bhartṛhari gives utmost importance to *śabda*. For *śabda* is not a mere means or avenue to truth or reality but it is the truth and reality. The awareness of this fact leads one to the realization of the Supreme Word. And this realization happens in a stage of *pratibhā* (intuition) which may very well resemble the *anubhava* of Śaṇkara's thought. Through his careful analysis of *śabda*, Bhartṛhari has thrown much light on the primordial mystery of language. Language is not merely an instrument of human communication but an indicator of the hidden Word which activates language as a meaning bearing medium of human life. That is why he emphasizes the need for purifying one's speech. As

---

[105] G. SASRTI, *Philosophy of Word and Meaning*, 9-12.

Bhartṛhari says in his *Nītiśataka*, right speech (*vāṇi*) alone adds to the handsomeness of one's personality more than anything else. While other things perish the gracefulness of speech lasts forever.[106] Bhartṛhari's conception of the levels of language is no doubt unparallel. Bhartṛhari talks of the levels of language in such a way that each level expresses the meaning of the whole. The inner meaning is deep within human consciousness and can only be imperfectly symbolized at the surface levels. Scholars like Manickam, see a tremendous theological significance of the emergence of world from the power of *śabda* as explained by Bhartṛhari.[107]

In this way Bhartṛhari has succeeded in setting a philosophical foundation for *Vyākaraṇa*. His *Vākyapadīya* is a *Vyākaraṇa Darśana*. In a word, he has made an attempt to present an integral vision of the Real in the framework of a Philosophy of Grammar. This approach depends for its validity upon the pre-supposition that the Real is a self luminous Truth which needs to be discovered by every speaker and in every speech something of the Real is revealed to the hearer. In presenting such a vision of a Philosophy of Grammar and Language Bhartṛhari has shown his own style and originality which made his Philosophy one of the richest varieties of linguistic philosophies the world has produced so far.

However, there are certain inadequacies in Bhartṛhari's

---

[106] *keyūrāṇi na bhūṣayanti puruṣam hārā na candrojjvalā /*
na snānam na vilepanam na kusumam nāalamkṛtā mūrdhajāḥ /
vaṇyekā samalamkaroti puruṣam yā samskṛtā dhāryate
kṣīyante' khila bhūṣṇāni satatam vāgbhūṣaṇam bhūṣaṇam // Nītiśataka. 15. W.L.S. PANSIKAR, ed., *Subhāshita Triśati of Bhartṛhari*,(Bombay: Pandurang Jawji 1925), 12.

[107] T. MANICKAM., "Cross Cultural Hermeneutics: The Patterns of Jaimini, Bhartṛhari and Śaṇkarāchārya," *Indian theological Studies* 21/3-4 (1984) 250-267.

system. On many points, there is not enough information. For example, nowhere he says anything about the transformation of knowledge into language, forms and vice-versa. Again he is silent on the method of learning the meanings of words. Although he visualized Grammar as a means of attaining *mokṣa* he does not describe the stages of this spiritual ascent. Yet his main purpose in *Vākyapadīya* seems to be rolling up the curtain on the stage of human communication so that the communicants may have a *darśan* of the hidden mystery of all meanings behind all words of human dialogue and performance on the stage of social communication. Creditable as it is, Bhartṛhari's contribution to the Philosophy of Language is original and indigenous to the Indian heritage of philosophy.

CHAPTER THREE

HEIDEGGER'S CONCEPT OF REALITY

1. INTRODUCTION

There is only one philosophical question that interests Heidegger, the question about Being[1] and its truth. This is the "one star"- the only - that remains constant along the way.[2] All his other problems are either concerned with the same problem of Being approached differently and revealed in different aspects, or they are concerned with different paths leading to this basic, abiding experience. Hence for Heidegger, the history of philosophy is the history of the interpretations of the Being question. All philosophy in the West, he thinks is ultimately metaphysical in nature. Being has been the subject of every metaphysical account of beings from Plato to Nietzsche. Metaphysics somehow missed the basic distinction between the question about particular beings (*Seiende*) and the question about Being as such (*Sein*); instead it kept on describing the nature and hierarchy of beings forgetting the question about the meaning of

---

[1]The word 'Being,' the basic concept in Heidegger's philosophy, will be written with a capital 'B' in order to distinguish it from the word 'being' which expresses a particular being or 'being' which is the participle of the verb 'to be.' 'Being' will correspond to the German *Sein* and 'being' to the German *Seiendes.*

[2]M. HEIDEGGER, "The Thinker as Poet" in *Poetry, Language, Thought,* tr., A. HOFSTADTER (New York: Harper & Row 1971), 4.

Being, about the "is." Traditional metaphysics, Heidegger claims, simply identifies Being as such with *a* particular being.

What Being means, Heidegger says, has fallen into oblivion (*Seinsvergessenheit*), precisely because the difference between Being and beings which he calls ontological difference[3] has been concealed. Being is everywhere reduced to the proportions of something entitative, of some being or other. For Heidegger Being is not *a* being though it is concealed in it as the Being of that very being. It is human thinking that unveils this Being. In the thought Being (*Seinsdenken*) practiced by Heidegger Being breaks forth as presence by which the being shows itself from itself. This experience of thinking leads to a new understanding of the Being question. Being is presence, the unending process of unveiling of Being from being. Heidegger, as the philosopher of

---

[3]Heidegger often said that "the forgetfulness of Being is the forgetfulness of the difference between Being and beings." M. HEIDEGGER, *Kant and the Problem of Metaphysics*, tr., J.S. CHURCHILL (Bloomington: Indiana University Press 1962), 243. This difference was forgotten because of an ambiguity intrinsic in the expression commonly used to refer to beings as the subject matter of metaphysics. The expression "being as being," *to on hei on* was used to refer to the whole of beings considered in terms of what makes them be, their "beingness" (*ousia*). The beingness of the whole of beings can mean just the common denominator of all beings, "being in general." But the beingness of beings can also refer to the ultimate ground which lets the totality of beings be. According to Richardson grammatically *ὄν (on)* is a participle and as such may be used either as a noun or as an adjective with a verbal sense. Taken as a noun it means that which is, i.e., a being (*Seiendes)*; and when taken as a verbal adjective (*seiend*) it designates that process by which a being (as noun) "is," i.e., its Being (*Sein*). The word thus, comporting both senses is intrinsically ambivalent and can mean either Being or beings, or both. This ambivalence points to a difference between Being and beings which Heidegger calls the "ontological difference." W.J. RICHARDSON, *Heidegger:Through Phenomenology to Thought* (Hague: Martinus Nijhoff 1963), 10-12. Hereafter W.J. RICHARDSON, *Phenomenology to Thought.*

Being in an epoch that is marked by the forgetfulness of Being, by the thoughtlessness about the root of beings in Being explains his own philosophy as a rethinking of the traditional metaphysics of being.[4] The new way of meditative thinking demands to the summons of Being and is realized through a dialogue with the historical roots of the entire culture of the West where Being always revealed concealingly.

Heidegger launches this search for Being in BT with an analysis guided consistently, carefully, and exhaustively toward the problem of Being. He continues his Being question in the post-*Being and Time* works, which are a multitude of scattered, often small, but profound approaches revealing Being in diverse contexts and aspects. All of them have one and the same basic character - the striving to reveal Being in its profound meaning.

We make an attempt to delve into Heidegger's Being question basing our study upon his most significant works in this regard. The methodology of this study is governed by the demands of its task: it is geared to analyse and thus to understand the question of Being from within Heidegger's thinking, emphasing more on his later thought. However, we follow the thematic unity rather than a mere chronological continuity of the writings.

2. THE BEING QUESTION

Being (*das Seiende*) as being is always sought for in Metaphysics. It has tried to understand being in the light of its Being. But how this very Being of being reveals itself in being is not sufficiently investigated nor seriously taken into consideration. It remains hidden. Because the difference between Being and being is forgotten by thinking Being as a being, that which throws being into beingness, itself comes to be considered as a being.

---

[4]M. HEIDEGGER "Only a God Can save Us: *Der Spiegel*'s Interview with Martin Heidegger," tr., M. P. ALTER & J. D. CAPUTO, *Philosophy Today*, 20 (1976) 281.

Instead of considering being in the light of Being, traditional philosophy thinks of them in respect to its relation to a subject. In respect to a subject, things are objects. Beings as objects can then only be thought of by forgetting Being itself. Such a philosophical attitude which thinks of beings as objects without considering Being itself, according to Heidegger is subjectivism. Metaphysics, instead of as investigating the objectivity of beings, has become subjectivism.[5]

The question of Being, according to this traditional metaphysical understanding was being asked as a question about *a* being, as a question regarding the beingness of any particular being as such; it was not a question concerned with Being as such. The question of the Being of beings, continues Heidegger, is a twofold question: 1. What is being in general? 2. Which one of the particular beings is the highest being (the supreme being among them), and how is it the highest? The twofoldness of the meaning of the question of the Being of being (*Sein des Seienden*), that is, of the question "What is being?"(*Was ist das Seiende*?), accounts for the nature of onto-theological questioning. The described Being-question is a question now turns out to be ontology (being in general) and of theology (the highest being). The metaphysical Being-question, therefore, is an onto-theological question.[6] As

---

[5]In regarding traditional philosophy as subjectivism, Heidegger takes a rather negative stand. He seems to be positively related only to the early Greek philosophers and to the meditating poet Hölderlin. Anaximander, Heraclitus, and Parmenides were true thinkers of Being. Heidegger considers this beginning of Western thought as the greatest phase in the history of Western philosophy which ended with Aristotle. There already was no philosophy of Being but merely of beings. Being was thought of solely as 'beingness' of beings (the entity of beings) but not as presencing of the realm in which beings present and thus appear.

[6]Heidegger rejects the onto-theological character of metaphysics (the identification of Being with a being, with God; the forgetting of the ontological difference) not on the ground of some atheism but in the light of

Richardson points out it is in Plato metaphysics in the traditional sense takes its rise, for it is he who first conceives of thinking Being as a going "beyond" the beings to their being-ness, which he conceives as their what-ness, their Idea.[7]

Heidegger is a philosopher who tries to think Being instead of the 'beingness' of beings. By questioning the nature of metaphysics Heidegger tries to transcend metaphysics and leads the mind out of the subject-object level into the level of Being, i.e., into a transmetaphysical level. His thinking begins with a question which has remained uniquely and exclusively the guiding question of western metaphysics. It is the question which Aristotle formulated at the beginning of the classical Greek philosophy: *ti to on*? What is a being in its Being? ("*Was ist das Seiende, das Seiende in seinem Sein*?"). The "first" science in philosophy has the task of grasping beings as being, thus of determining their Being.

The problem of Being arose as soon as Heidegger began to meditate with Brentano (1838-1917) the meaning of the word "being" for Aristotle.[8] In 1963 he called Brentano's book "the chief help and guide of my first awkward attempts to penetrate into philosophy," adding, "the following question concerned me in quite a vague manner: If being is predicated with manifold significance, then what is its leading, fundamental signification? What does Being mean?"[9]

Thus, Heidegger became fascinated by "is," the little word that applies to everything, without detriment to the marvellous unity

---

a new way of thinking, the yet unthought unity of the essence of metaphysics.

[7]W.J. RICHARDSON, *Phenomenology to Thought,* 104.

[8]Heidegger's Being question initiates as he claims with Franz Brentano's dissertation on "being" in Aristotle. (*On the Manifold Meaning of Being according to Aristotle*, 1862).

[9]M. HEIDEGGER, "My way to Phenomenology," in *On Time and Being*, tr., J. STAMBAUGH (New York: Harper & Row 1972), 74.

of itself. Yet what of this unity? This must be Being itself, that which renders all beings possible. Well, then, what about Being? What meaning does it have? If it is true, as Aristotle says, that the function of metaphysics is to ask "what are beings as beings?, then, on the supposition that Being gives beings their "is," should we not first ask about Being itself? Certain peculiarities of the Greek language also favoured the development of the "science of being," as Aristotle saw metaphysics. As Frede observes, even in pre-philosophical Greek it was quite common to refer to "beings," to "what there is," both in the sense of "things" and in the sense of what we would call "states of affairs." The fact that there is a clear linguistic distinction between "being," *ta onta,* referred to by the participle with the definite article, the verb "to be," *einai*, and the abstract noun "being," *ousia* (the nature of beings), makes the development of such a philosophical discipline much more natural.[10] It became natural to ask whether there is a unitary meaningful concept that demarcates the realm of Being as such. Plato raises the question in the *Sophist*: whether there is a unified meaning of being that accrues to all beings. It seems Heidegger was well aware of Plato's struggle with this problem, since he used the passage in the *Sophist* as his point of departure in BT.[11]

Whatever Plato may have thought about the "unity of being," it was the Aristotelian doctrine of a manifold of meanings of being that came to dominate the history of Western metaphysics. While the question of Being still occupied a certain position in the investigations of Plato and Aristotle, it was later forgotten. Hence

---

[10]D. FREDE, "The question of being: Heidegger's Project," in *The Cambridge Companion to Heidegger*, ed., C. GUIGNON (Cambridge: University Press 1995), 43.

[11]BT. 19. For a comprehensive discussion of the different meanings of "being" and the importance of the distinction between the copulative, existential, and veridical senses of "is" for the development of philosophy, see C.H. KAHN, *The Verb 'Be' in Ancient Greek* (Dordrecht: Reidel 1973).

Heidegger sees himself forced, to reawaken an understanding of the question of the meaning of Being. And he takes up the question of what the expression "being" (*seiend*) actually means in his master work BT.[12]

In the very first paragraphs of BT Heidegger speaks of the "ontic and ontological priority of the Being question." Though it is usually forgotten, this has a twofold meaning. It means that in philosophy the question about Being deserves priority above all other questions: "Being is the genuine theme, the only theme of philosophy."[13] It also means, however, that, as a question, the question about Being is the being *par excellence* which must serve as the point of departure for the question about the Being of beings.[14] In order to think Being, metaphysics has to be overcome. By not taking metaphysical principles for granted, but by trying to think them anew, metaphysics can be transcended.[15] In this sense Heidegger's thinking is a return to Being. Such a procedure tending to go backwards more deeply than traditional philosophy ever went, is called 'destruction'[16] by Heidegger.

The method Heidegger chooses for his philosophical investigation is Phenomenology. The question of the meaning of Being is one that must be treated phenomenologically.

---

[12]BT. 19.

[13]M. HEIDEGGER, *Die Grundprobleme der Phänomenologie*, GA 24 (Frankfurt am Main: Klostermann 1975), 15.

[14]GA 24. 296.

[15]Even though metaphysics ignores Being, Heidegger does not suggest its rejection. The forgetfulness of Being is a characterization and not a devaluation of metaphysics.

[16]The German word *"Destruktion,"* as Heidegger uses it, is not to be understood in the sense of a destroying or as a deconstruction of tradition. It means the overcoming of the old metaphysics and the discovery of a new metaphysics, of a new understanding of Being. With his 'destructive' analysis of the history of philosophy, Heidegger leads the basic problems of past philosophers back beyond their starting points into the realm of Being itself.

Phenomenology according to Heidegger is 'to let anything manifest itself in the very way it manifests itself from itself.' This is the impartial approach to the things themselves.[17] Further phenomenology for Heidegger is a way to what is to be the theme of ontology.[18] From the very outset of his procedure he says that Phenomenology and ontology pertain to the one and the same philosophical discipline and that ontology is possible only as Phenomenology.[19] This phenomenological ontology aims at the very manifestation of Being itself, revealing and concealing and concealing and revealing as it is.

In BT Heidegger formulates the question of Being in the form of a fundamental ontology,[20] for a better understanding concerning beings as beings and concerning the Being of beings. The fundamental ontology involves a twofold task[21] of the analysis of *Dasein*[22] *(Analytik des Daseins*) and a structural analysis of the already existing ontology (*Destruktion der bisherigen Ontologie*). The fundamental ontology of BT is concerned with that which metaphysics has left unthought up to

---

[17]BT. 50.

[18]BT. 62.

[19]According to GA 24 the method of ontology is phenomenology. Phenomenology itself is characterized by three basic components: the phenomenological reduction, the phenomenological construction and the phenomenological destruction. (P. 31).

[20]Heidegger distinguishes metaphysics from ontology by considering ontology as the proper study of Being and metaphysics as an investigation of beings. Since ontology has been often used as a synonym for metaphysics, he calls his own study of Being 'fundamental ontology.'

[21]BT. 36 ff.

[22]The word Dasein is very important in Heidegger's philosophy and is a special German expression with many indications and connotations. Literally it would mean There-being (*Da-Sein*) which would signify the relatedness of human reality to Being as well as the sense of presence, the mode of "to be" that distinguishes it from all other beings.

this day and thus, with the ground (*Grund*) and the abyss (*Abgrund*) of all metaphysics and ontology.[23] The structure of the first part of this work is determined by the attempt to tear thought away from its orientation toward things and to lead it back to its own most self, the Dasein, and its temporality so that it becomes possible to give a determination of the meaning of Being.

Heidegger's purpose is to lay the groundwork for metaphysics by seeking to discern that ontological structure of man which is the source of his natural tendency to metaphysical thinking. Since metaphysics deals with Being structures and the metaphysician is intrincically finite, the problem becomes an effort to explain the relation between Being and finitude. The comprehension of being as such is finite and BT attempts to discern phenomenologically the ultimate sense of this finite comprehension of Being. Though finite in comprehension, it is an earnest attempt to all beings by virtue of its orientation to Being. Thus Dasein is, not an entity enclosed within itself but essentially an open-ness towards Being which again is characterized by its own finitude and realizes itself in time only as a process that comes-to-pass. Therefore Heidegger calls man a 'clearing,'[24] the place or the 'there' where being is brought to light.

Heidegger makes a phenomenological analysis of Dasein, the There-being. It is a special sort of phenomenology, for in discerning the sense of There-being, There-being itself is engaged. Hence, if we consider in all its purity this hermeneutic

---

[23]In later years, Heidegger seems to have become increasingly skeptical about the enterprise of a fundamental ontology that "lays bare" the structure of being as such, since this now seems to him a kind of "foundational" enterprise that reeks of metaphysics, the project of establishing an ultimate basis for all things. To make human understanding the key to such a transcendental investigation, somehow suggests that we are in *control* of the "Being" of all beings.

[24]BT. 171.

interpretation by There-being of its self, we may say that phenomenology comes to its fullness when There-being achieves its authenticity. The phenomenology of BT, then, culminates in the moment of re-solve in which knowing and opting become one and the same act of the There-Being. Thus Heidegger strives constantly to overcome the dichotomies that have been so typical of Western thinking and to show that they grew out of what was originally a unity. He asserts the essential link between thinking and opting, Dasein and Being.

3. BEING AND DASEIN

As we have seen, the discovery of Being according to the phenomenology of BT is made possible by the right way of asking, by an authentic questioning. The questioned is not *a* being, but Being itself. How do we come to the discovery of Being? Through the understanding of There-being (Dasein).[25] Because Dasein is characterized by the potentiality of asking the Being-question, the elaboration of the question about the meaning of Being (*Sinn von Sein*) requires a previous explication *a* particularly privileged being, i.e., Dasein from the point of view of its own Being.[26] Dasein, man as a privileged being, always has a relation to its Being. Dasein is privileged because Da-sein is gifted with awareness of its own Being. Therefore he is the *Da* of *Sein*, i.e., the *Da* where Being (*Sein*) shines forth. Hence Dasein's "essence" lies in its existence, in its "drive-to-be" (*Zu-sein*).[27] This is what distinguishes Dasein from other beings as mere entities (i.e., *Vorhandenes*).

---

[25]Even though There-being is not man as such but that "place" where the essence of man has its abiding ground, i.e., the ultimate source out of which man comes-to-presence as man, Heidegger seems to use these terms often interchangeably.

[26]BT. 27.

[27]BT. 67.

The Being of beings is not something that is just "out there" all by itself, or even "out there" in beings; rather, it implies the meaningful relatedness and the intelligible presentness of things to and for man understood as Dasein. The probing of the problem of Being means then that the problem of man must be settled first. To Heidegger, phenomenology thus devolves into an analytic of the Dasein, the being who is "there." So Heidegger considers the question of Dasein as a prelude to the question of Being.

There is, says Heidegger, a remarkable relation both forward and backward between what is questioned (Being) and the questioning itself as a mode of Being of a being (Dasein). Heidegger sharpens this point in "The Way back into the ground of Metaphysics": "To characterize with single term both the involvement of Being in human nature and the essential relation of man to the openness ('there') of Being as such, the name of 'being there'(Dasein) was chosen for that sphere of Being in which man as man stands."[28] In short, man as a questioning being is the way to the questioning of Being, leading, finally, to Being as the questioned. Therefore, Being for Heidegger is always Being as it enters into Dasein's understanding of Being. There is Being, Heidegger says, only insofar as it is understood by Dasein.[29] Being is always thought in terms of a horizon which is projected by Dasein. By Dasein, Heidegger understands man as the "there," i.e., as the place of the disclosure of Being.

It is Dasein which raises the question of Being. Dasein can ask about Being because it is distinct from other beings in that, in its Being, Dasein is aware of and concerned about this very Being. It is ontically and ontologically prior to all other beings. It is ontically prior because it is a being (*Seiendes*) whose Being (*Sein*)

---

[28]M. HEIDEGGER, "The way back into the ground of Metaphysics," tr., W. KAUFMANN in *Existentialism from Dostoevsky to Satre*, ed., W. KAUFMANN (NY: World-Meridian 1956), 213.

[29]BT. 255.

has the determinative character of existence. It is also ontologically prior to all other beings because it is endowed with the privilege of understanding Being.[30] Being the *Da* (there) of its own *Sein* (Being) man is the Da-sein.[31] But, because the Being of Dasein is available only in a projected world, the *Da* is the world itself and Da-sein is the Being-in-the-World (*In-der-Welt-sein*).[32]

Since the essence of Dasein lies in "ek-sistence,"[33] in its being-able-to-be (*Sein-können*), understanding Being is a characteristic of the Being of Dasein. Thus, Dasein has not only an ontic priority- as being among beings- but also an ontological priority: Dasein is in itself ontological: it has an understanding of Being. The comprehension of Being therefore, is not an abstract idea but a movement, a continuous going from existence to Being and from Being back to existence. The reality of this movement is not a fixture of the teleology of things (*Vorhandenes*); it is the foundational dynamism of the openness of existence (*Dasein*) to Being (*Sein*), the phenomenon of primordial transcendence

---

[30]BT. 34.

[31]BT. 171.

[32]BT. 182.

[33]Heidegger makes use of the term ek-sistence in contrast to existence, in order to point out the nature of man's existence. Because Being is ecstatic in nature, it "is the ecstatic 'standing-out' of man in to the truth of Being." M. HEIDEGGER, *Letter on Humanism* in *Basic Writings*, ed., D.F. KRELL (New York: Harper Collins 1993), 228. So man's very existence is an ek-sistence, namely, a going out of himself. This going out signifies the fundamental character of man's ek-sistence as openness to Being. Being offers itself to the openness of man and only in this openness of man, in this *Da* of Dasein could Being essentialize itself as Being. According to Vycinas Heidegger modifies the way of writing this word to accentuate the existence as a relation to Being instead of to one's own being. V. VYCINAS, *Earth and Gods* (Hague: Martinus Nijhoff 1961), 72. Hereafter V. VYCINAS, *Earth and Gods.* To ek-sist means not only be what one is but also be the possibility of standing out into one's beyondness.

(*Urtranszendenz, ursprüngliche Transzendenz*).[34] There-being comprehends the Being of beings. By this very fact, There-being passes beyond or transcends beings to the Being-process as such.

According to Heidegger the metaphysical conception of man as a rational animal fails to take into account the relationship of man's essence to Being. Man's ek-sistence is ek-static open-ness to Being. "...Such standing in the clearing of Being is what I call the ek-sistence of man..."[35] Dasein is held in releasement to Being by Being itself. This affirmative mode of pure openness to Being is *ek-stasis*. This ek-static standing of Dasein is also a standing out in the truth of Being.[36] Being offers itself to the openness of man and only in this openness of man, in this *Da* of Dasein could Being essentialize itself as Being. *Da* is the illumination of Being.[37] According to Vycinas Dasein for Heidegger is primarily the presence of Being in concrete life and situations. Man is the only being who has an understanding of Being, therefore, only for man can Being be present. Man is the place of the presence of Being; he is this presence. Man is Dasein. Dasein is not identical with empirical man, but is rather his essence.[38] Being is not a statically unchangeable whole: it constantly brings itself forward into openness, into the *Da* of Being. And man is not a closed-in-itself being: he stands out in the openness of Being and is the carrier of this openness - he 'ek-sists.'

Being can only be revealed in Dasein. Therefore, as long as

---

[34]G. KOVACS, *The Question of God in Heidegger's Phenomenology* (Evanston: Northwestern University Press 1990), 217. Hereafter G. KOVACS, *The Question of God.*

[35]LH. 228. Heidegger insists again on the difference between ek-sistence and the *existentia* of the tradition, for he attempts to get to a point prior to the essence-existence dichotomy.

[36]LH. 229.

[37]LH. 241.

[38]G. VYCINAS, *Earth and Gods*, 24-25.

there is a Dasein, Being can be revealed. On the other hand, man is man because Being breaks into openness which is the essence of man. The *Da* in Dasein is Being itself revealed or brought out of concealment into disclosure. Since disclosure in Heideggerian thinking is truth, Dasein is the truth of Being. Dasein is the place of the truth of Being. Without Dasein Being is not Being in the sense that it is not revealed; it does not come out of concealment into revelation, which means there cannot be any truth of Being. Dasein and Being can never be considered as two separate beings.[39] However as Richardson asserts, Dasein is not the source of its own Being but rather finds itself as an already existing fact, i.e., immersed in its original situation as a comprehension of the Being of beings.[40] Heidegger describes Dasein's non-mastery of its own origin and its referential dependence on other beings with the term "thrown-ness" (*Geworfenheit*),[41] pointing to the matter of fact character of human existence.

Further, Heidegger conceives man as a being-to-death or a being-to-nothing, i.e., a being whose existence is marked by having to realize himself in the face of death or the nothing by which he is threatened. In so doing, he becomes conscious of himself as the torchbearer of true Being, which, through man's thinking and language, illuminates the world. Man is the project (*Entwurf*) or illumination of Being (*Lichtung des Seins*). As such he leads an authentic existence. In this authentic existence, the truth of Being is revealed. In this revelation or unveiling of truth, man experiences the metaphysical, namely, Being as the fountain

---

[39]In BT Heidegger insists on the fact that There-being is not to be understood as a subject, although it is unquestionably a self - a pre-subjective self that precedes the dichotomy of subject and object and renders this dichotomy possible, i.e., it is transcendence. Heidegger conceives man on a level deeper than that of the traditional essence-existence dichotomy. He interprets man purely out of his relationship to Being - as transcendence, as ek-sistence.

[40]W.J. RICHARDSON, *Phenomenology to Thought*, 37.

[41]BT. 174; 219.

of his thinking and culture. [42]

Nevertheless, Man alienates himself, from his authentic existence as a being-to-death and acts in an unauthentic way. This happens because man imagines that he, through constitutive metaphysics, comprehends Being and thus controls the world. In reality, however, Being is incomprehensible. It is not man that disposes of Being, but it is Being that disposes of man.[43] Thus, human being's relation to Being, in the final analysis, is founded on the relation of Being to the human being. Thinking is not causing or bringing about this relation; it is concerned with the relation of Being to the human being; it "accomplishes (*vollbringt*; brings full) the relation of Being to the essence of man" as "something handed over to it from Being."[44] The essential relationship of the human being to Being is founded on "the relation of Being to the essence of the human being."[45] The human being is concerned with grounding the truth of Being, with the shining forth of being in the light of Being because he "is 'thrown' from Being itself into the truth of Being" in order to exist (ek-sist) in this way;[46] man is not the ruler but the guardian of Being; his way of being is bound up with the destiny of Being.

There-being is the "There" (presence) of Being; human existence is, in the final analysis, the openness to Being, "the standing in the lighting-process of Being" (*das Stehen in der Lichtung des Seins*).[47] The human being dwells in the light of Being; the lighting-process (Being) "grants the nearness to

---

[42]G.A. RAUCHE, "Back to Metaphysics," *Man and World* 10 (1977), 399-400.

[43]G.A. RAUCHE, "Back to Metaphysics," 399-400.

[44]LH. 217.

[45]LH. 228.

[46]LH. 231.

[47]LH. 229.

Being."[48]Man "is" *nothing* but the free space for the presence of Being. Such is his primary definition, which in Heidegger's eyes is not restrictive:

> What is distinctive about man resides in the fact that, as the one whose essence it is to think, open towards Being, he is placed before the latter, remains related to Being and thus corresponds to it. Man properly is this relation of correspondence, and he is only this. 'Only': this does not mean a limitation, but an overabundance. In man there prevails a belonging to Being, a belonging which listens to Being, because it is appropriated over to being.[49]

The essence of man is to depend on Being, to correspond to it and to belong to it as proper to it (*eigen*). Immediately following this passage, the text of *Identity and Difference* seems to introduce a reservation and make apparent an equal dependency of Being with respect to man. Indeed, would Being come to presence without man? Apparently not. "Being essences and endures only insofar as it attains to man through the claim of its address. For it is man, open for Being, who first lets Being arrive as presencing."Yet there is an immediate corrective: "this does not at all mean that Being is first posited by man and only through him."[50] The inequality of the relation as Harr observes is that man does not posit himself, but is put in the world by Being; Being, on the other hand, is never something fabricated by man. Being produces man; man cannot produce Being, nor can he produce himself.[51]

Man finds himself situated at the intersection of two movements: his own that transports him into Being, and the movement via which Being retracts human transcendence into itself. "Man stands within the relation of Being itself to him,

---

[48]LH. 223.

[49]M. HEIDEGGER, *Identity and Difference*, tr., J. STAMBAUGH (New York: Harper & Row 1974), 31.

[50]ID. 31.

[51]M. HARR, *Heidegger and the Essence of Man* (Albany: State University of New York 1993), 65. Hereafter M. HARR, *Essence of Man.*

man."[52] Man himself does not decide about this relational essence defining him, rather his relation to Being responds or corresponds to Being's relation to him, which is always prior. Man is in a situation of receptivity, if not passivity. He is above all *correspondence*: *Entsprechung*, which for Heidegger also means "speaking in return," the response to silent "address" of Being.[53]

The relationship between Being and man may be compared to a summoning - hearing relationship. Any presence (Being) by the very fact that it is a presence, is a presence to the essence of man and in so far as presence is a call, it summons the essence of man.[54] In order to exist, man must carefully attend to the silent summoning of Being and attending the call preserve it in truth. As the preserver of the truth of Being, man is the shepherd of Being.[55] The essence of man thus, lies in his openness to the address of Being. In this openness man releases the "thing" to be what it is. This openness to the address of Being is therefore the "letting-lie-forth"(*vorliegenlassen*) by which the thing is left to rest on its own grounds. By means of this openness, which belongs to his very essence man is admitted into a new region, the region in which "reasons" are neither demanded nor supplied, in which things stand forth and emerge out of their own grounds, free from the categories of metaphysical thought.

From the outset, Heidegger had seen clearly that the relationship between Being and man cannot be thought of in terms of the subject-object relation. From the very beginning it was also clear that man's relationship to Being is constitutive for man's essence. But whereas in the earlier treatises, as far as the discourse about this relationship is concerned, the focus was

---

[52]M. HARR, *Essence of Man,* 69.

[53]M. HARR, *Essence of Man*, 70.

[54]F.J. VADAKETHALA, *Discovery of Being* (Bangalore: Dharmaram Publications 1970), 69. Hereafter F.J. VADAKETHALA, *Discovery of Being*.

[55]LH. 234.

primarily on man's *Da* as transcendence, in the later treatises the focus shifted for the most part to that toward which Dasein transcends beings, i.e., Being itself. It is then explicitly claimed that Being as coming-to-presence is a demand which summons and hails Dasein to its essence. Thus the very essence of Dasein is such that it implies that it is continuously in a state of listening, because its essence belongs to the summoning and evoking hail of Being, i.e., because it belongs to its coming-to-presence itself.[56]

We have noted elsewhere that Being for Heidegger is fundamentally a lighting-process by which beings are lit up as what they "are." Now in this coming-to-pass, Being maintains the initiative. It is Being that gives itself unto the There with the There and, imparting itself thus, continues to dominate its There. This is the mittence of Being. Yet mittence takes place for the sake of Being, and the There, thus brought-to-pass, sustains the process, guards it, stands watch over it so that in the light of Being beings may appear non-concealed- such is the concern of the There. Now the Being that comes-to-pass is also equally a non-Being.

4. BEING AND NON-BEING

In the traditional sense anything that is not being (*Seiendes*) is logically termed as nothing. Traditional metaphysics speaks of nothing "in the quite ambiguous proposition: *ex nihilo nihil fit*,

---

[56]For a Comparative understanding of the views of Heidegger and Eckhart in this regard, see J.D. CAPUTO, *The Mystical Elements in Heidegger's Thought* (New York: Fordham University Press 1986). Hereafter J.D. CAPUTO, *Mystical Elements.* What is essential about Being for Heidegger is that it differentiates itself from beings, is other than beings. And if man is ever to gain access to such Being, it can only be by way of realizing and opening himself to this "difference." There is an "ontological Difference" in Eckhart as well as in Heidegger, and in understanding this Difference lies one key to the unity - or belonging together- of Being and man in both thinkers.

nothing comes from nothing."[57] This proposition interprets 'nothing' from the point of view of the metaphysical conception of beings, and thus it leaves unasked the whole question of nothing and Being as such. Therefore, Heidegger insists, the metaphysical notion of nothing speaks to us more about the metaphysical conception of being(s) than about the question of nothing. For we tend to treat nothing as if it were a "being" which "is," "yet from this," Heidegger adds, "it is precisely differentiated" (*unterschieden*).[58] The nothing is what is "cut off" and "separated" from beings. It is no thing, not any thing at all. From the point of view of the logician, nothing is a truth-functional concept. That is, it arises by taking the totality of what is and then negating it. Nothing is a concept which arises out of the logical function of negation. But for Heidegger the nothing is something experienced, encountered, something forceful which we "run into." The nothing for Heidegger is reached by a "fundamental experience" (*Grunderfahrung*).[59]

Now Heidegger thinks that there is something which manifests itself in and with Being but not Being itself, which he calls 'Nothing.' Nothing is said to be nothing precisely because it is No-thing.[60] Being, as that by which all things are, is not a being, nor the collection of beings- it "is" Non-being (*das Nichts*). Therefore, Being is to be understood in its contrast to Non-Being.

---

[57]M. HEIDEGGER, *What is Metaphysics*, tr., R.F.C. HULL & A. CRICK, in W. BROCK, *Existence and Being*, 344.

[58]WM. 329.

[59]WM. 333.

[60]Heidegger makes use of the word *Nichts* in the sense of *das Nicht-Seiende* (WM. 329). The English equivalent for *nichts* is 'Nothing,' which when written as No-thing indicates to some extent the intended meaning of *das Nichts*, that is, *das Nicht-Seiende*: No-thing; Non-being. For a detailed study of the problem of Nothing and Negativity, see N. SPROKEL, *Das Nichts und die Negativität nach Martin Heidegger* (Roma: Pontificiae Universitatis Gregorianae 1972).

Everywhere and every time Being is questioned, Nothing also is questioned with it. Hence the fundamental question assumes the form: "Why are there beings rather than Nothing?"[61] In analyzing the question of non-being Richardson makes the following comment: 'Non-being is not an Absolute Nothing but Being itself, considered as Other than beings. "...This completely Other to all beings is Non-being. ..." and "... Non-being as the Other to beings is the veil of Being. ..."[62] It is clear then that for Heidegger the Nothing is not something apart from beings. It is not the "counter- concept" to the Being but rather belongs to the essence of the Being.[63] Indeed the Nothing is the same as Being itself: 'in the Being of beings, the negating of the Nothing *(das Nichten des Nichts)* come to pass.'[64] By Nothing Heidegger does not mean the simple denial of what is, the negation of all beings, but Being itself, taken as what is other than any being (*Andere zu allem Seienden*). The negating of the Nothing is the Being of beings: "...that which never and nowhere is a being reveals itself as that which differentiates itself from all beings which we call Being."[65]

Again without this Nothing, which is Being, every being would fall into beinglessness (*Seinlosigkeit*), for "a being never is without Being."[66] Since Being is in no way a being, but is rather not-a-being, it can only take place as nothingness in the reality.[67]

---

[61]M. HEIDEGGER, *Introduction to Metaphysics*, tr., R. MANHEIM (New Haven: Yale University Press 1987), 1.

[62]W.J. RICHARDSON, *Phenomenology to Thought*, 474.

[63]WM. 340.

[64]WM. 340.

[65]WM. 353.

[66]WM. 354.

[67]To signify Being as not-a- being and as not opposed to nothingness, in one of his later works *Zur Seinsfrage*, Heidegger writes the word Being with crossmarks. Crossmarks indicate as Vycinas points out that Being does not exclude nothingness; it implies it in itself. Being is not a thing; it is not

Being and nothingness belong together; Being and nothingness are not outside of one another. Nothingness is never just a negated something. "Nothingness is prior to the 'no' and the negation."[68] Nothingness is not, but nevertheless it is real and belongs to the event of Being.

In Heidegger's view it is anxiety that reveals Nothing.[69] And it is the world itself before which Dasein is anxious. The world, which, however, is no-thing and hence appears itself as Nothing. It is the world in which Dasein has to find its meaningful existence and thus it is the very 'Being-in-the-world' of Dasein before which Dasein is anxious.[70] Because the 'Being-in-the-world' is the characteristic note of Dasein's existence, it is about its own existence that Dasein finds itself anxious.

Anxiety in so far as it shows the Nothing of Being makes Dasein conscious of its inauthentic state of existence and bringing this consciousness invites it to authentic existence. To exist authentically Dasein should understand the true nature of Being and to understand the true nature of Being Dasein has to understand itself, because it is only in Dasein that the true nature of Being is revealed. An untrue understanding of Being may leave Dasein in an inauthentic state of existence.[71] In anxiety we have

---

identical to the 'beingness' of the beings. Nevertheless it is not without the things or beings and their 'beingness.' V.VYCINAS, *Earth and Gods*, 106.

[68]WM. 331.

[69]WM. 335. Here anxiety is not understood in the sense of fear.

[70]BT. 230-232.

[71]The Being of Dasein is 'Being-in-the-world' (*In-der-Welt-sein*). But 'Being-in' of the 'Being-in-the-world' could realize itself in two different ways: authentic and inauthentic. F.J. Vadakethala explains it thus:To the authentic realization of the 'Being-in' of Dasein belongs the existential constitution of the 'Da' of Dasein which consists in the understanding of Dasein as 'Self-discovery'(*Befindlichkeit*); as understanding (*Verstehen*), and as discourse (*Rede*). To the inauthentic realization of the 'Being-in' of the Dasein belongs the every-day Being of the 'Da' of Dasein which consists in the manifestation

the real experience of Being[72] and the invitation to our authentic existence. As Vadakethala opines Nothing is only an attempt to understand Being through its opposite.[73]

In his later works, Heidegger continues to stress the negativity of Being; and it is in this connection that he states that the ontological meaning of Non-being implies that it can also be called Being. Further he understands Being not just as the Being of beings, but that it first and foremost is itself, the "other of beings," somehow antecedent to all beings.[74] It is unfolded with greater clarity when he articulates the ontological difference, i.e., the difference between beings and Being. In his thought, Heidegger is thus concerned with Being itself, taken in both its negative and its positive aspects. His entire philosophical effort is oriented toward an attempt at understanding the strange paradox that Being conceals itself in the beings which it makes manifest, that it hides by the very fact that it reveals.[75] When Heidegger thus speaks about Being as the process of non-concealment, he takes both the positive and negative aspects of this process as a unity. Richardson has described this process in the following statements: "Being as the process of non-concealment is that which permits beings to become non-concealed (positivity). Although the process is so permeated by 'not' that Being itself remains concealed (negativity). To think Being in its truth, then

---

of Dasein as ambiguity (*Zweideutigkeit*), as curiosity (*Neugier*), and as prattle (*Gerede*). These existentials of the every-day being (inauthentic existence) of Dasein constitute the 'falling' (*Verfallen*) of Dasein. F.J. VADAKETHALA, *Discovery of Being*, 58.

[72] BT. 229.

[73] F.J. VADAKETHALA, *Discovery of Being*, 62.

[74] WM. 360-61.

[75] For a better understanding of the concepts of revealing and concealing of Being in Heidegger's thought see, L.M. VAIL, *Heidegger and Ontological Difference* (New Jersey: The Pennsylvania University Press 1972), 25ff.

is to think it in terms of both positivity and negativity at once."[76] Therefore in the non-concealment of the Being - non-Being truth occurs.

5. BEING AND TRUTH.

Truth is a central theme within the context of Heidegger's project to retrieve the question of Being and to lay a foundation for the pursuit of that question. From the very beginning of his philosophical search he was concerned with the question of the truth of being. Heidegger argues that the correspondence between *intellectus* and *res*, posited in the traditional conception, lacks an adequate foundation.[77] The truth of Being is that center in which Being and Dasein (which has an understanding of Being) come together. Heidegger reflects upon this centre when, in two decades after the appearance of BT, he makes the problem of truth and Being the foremost theme of his thought.[78] The effort to

---

[76]W.J. RICHARDSON, *Phenomenology to Thought*, 9.

[77]Truth in the traditional sense is defined as *adaequatio rei et intellectus* (truth is the correspondence, the conformity, between the intellect and the thing). This definition may mean that truth is *adaequatio rei ad intellectum* (the conformity, the approximation of the thing to the intellect); it also can mean, however, that truth is *adaequatio intellectus ad rem* (the conformity of the intellect to the thing). According to the later interpretation, truth means the conformity of knowledge (of the intellect's act of judgment) to the thing (to the judged, to beings). In this case, then, truth means propositional, logical truth. According to the former interpretation truth means the conformity of things to knowledge. In this second case, then, truth means ontological truth. In both cases truth itself is defined as the rightness, as the correctness of the conformity. The place of truth is the judgment, the act of the intellect. See M. HEIDEGGER, *On the Essence of Truth*, trs., R.F.C. HULL & A. CRICK in W. BROCK, *Existence and Being*, 292-324.

[78]In fact, Being and truth have been linked together ever since the beginnings of philosophy in Greece. Aristotle defined first philosophy equally as the science of truth and as the science which investigates beings *qua* beings, i.e., with regard to their Being. Krell asserts that already in BT the double

ground metaphysics (fundamental ontology) began as a search to illuminate the intrinsic correlation between the Being process as such and the finitude of the being that comprehends it, i.e., There-being. The first step was to analyze There-being phenomenologically in order to find in the pre-ontic comprehension of Being some means of discerning the sense of Being. Subsequently the author has become more and more preoccupied with Being itself, but chiefly in terms of the problem of truth, since the sense of Being is its truth.[79]

Truth rather than as the *adequatio* of thought and thing, is thought of as a non-concealment. Heidegger uses the Greek term *orthotes* (truth as correctness) to refer to the notion of truth as propositional correctness to the already opened. But truth as the opening-concealing of the whole, grounded primordially in language and finding expression in various ways in the metaphysical tradition, Heidegger refers to by the Greek term

---

leitmotif of Heidegger's thought is the question of Being and the question of Truth. In section 44 of BT Heidegger writes, "Being and Truth 'are' equiprimordial." The double leitmotif of Being and Truth dominates literally all of Heidegger's later work. See D.F. KRELL, "On the Manifold Meaning of *Alētheia*: Brentano, Aristotle, Heidegger," *Research in Phenomenology* 5 (1975) 91-92.

[79]It is interesting to note Pöggeler's observation: 'We must differentiate precisely between the question of Being as concerned with the truth *of beings*, and the question of the truth of Being itself. Heidegger's question does not concern simply Being as the truth of beings, but concentrates entirely on the question of the truth of Being itself. Prior to the question which guides metaphysics (What is a being in its Being or truth?), he puts the fundamental question: What is Being itself in its truth? He tries to reveal the ground, never articulated as such, on which the metaphysical question already stands. In this way his question about the truth of Being itself is the return to the heretofore hidden ground of metaphysics.' O. PÖGGELER, "Metaphysics and the Topology of Being in Heidegger," in *Heidegger: Man and Thinker*, ed., T. SHEEHAN (Chicago: Precedent Publishing 1981) 178.

alētheia or truth as un-concealment, *a-lētheia.*[80] He stresses the *lēthe* as what is concealed through the manifestation, as the dimension of mystery, of the undisclosed to which we are, nonetheless, essentially related. For we are, at base, Dasein, the "there" of Being, the locus of the manifestation and concealment of the whole.[81] It is the happening of truth and in this happening prevails the temporality of Dasein and that time in which Being itself gives itself in its openness.

According to Kockelmans when Heidegger uses the expression "the meaning of Being," the word *meaning* (*Sinn*, "sense") means the non-concealment by which Being appears as itself.[82] Thus the expression "the meaning of Being" and the "the truth of Being" have the same meaning.[83] Foundational thought tries to meditate Being as the process of truth, i.e., the coming- to-pass of the lighting-process in beings. It is brought-to-pass by the nature of man conceived as ek-sistence, i.e., endowed with the prerogative, unique among beings, of an ecstatic open-ness unto the lighting-process of *alētheia.* In 1964 Heidegger cited as the persistent task of his thought the meaning of *alētheia* - no longer to be translated as "truth" but to be pondered as unconcealment or "the clearing that first grants Being and thinking and their presencing to and for each other."[84]

Truth understood in the radical sense of *alētheia* is thus

---

[80]*Alētheia,* literally means "to pluck something out of its concealment, to make manifest or reveal." BT. 262. It means the emergence into the open, the process by which things are brought into the clear, into the sphere of the un-concealed. Being true means being unconcealed or unveiled.

[81]M. HEIDEGGER, *On the Essence of Truth* in W. BROCK, *Existence and Being*, 312.

[82]J.J. KOCKELMANS, *On the Truth of Being* (Bloomington: Indiana University Press 1984), 54. Hereafter J.J.KOCKELMANS, *Truth of Being.*

[83]WM. 351-352; BT. 192-93.

[84]TB. 68.

unconcealment. Being as the process of non-concealment is that which permits beings to become non-concealed, although in the process Being itself remains concealed. The hiddenness of Being (in beings) may be understood in terms of a 'not' that contracts Being in beings and at the same time differentiates it from them. Since the function of Being is simply to en-light-en beings, then this contracting 'not' is intrinsic to its very nature. Richardson calls this 'not' character of Being "negativity" and the manifestive power that shines forth in beings as beings "positivity."[85] According to him when we comprehend this fusion of positivity and negativity into the unity of a single process, we begin to grasp what Heidegger understands by Being as the process of truth.

In disclosing itself in beings as their truth, Being withdraws behind the finitude of this revelation. Heidegger says: "... it hides itself in this [disclosure] and conceals itself [as] thus hiding." This concealing by Being of its own concealment is the "mystery" of Being "... according to which the truth of Being comes-to-presence."[86] The self-illuminating rise of Being comes-to-pass in such a way that Being itself actually hides in the background. The ascent of Being and beings is always from hiddenness, and keeps something of hiddenness always in it. Finally, it means that truth is not something manageable and fixed; rather, it is to be achieved, and it comes-to-pass little by little.[87] Here the Being in

---

[85]W.J. RICHARDSON, *Phenomenology to Thought,* 8-9. As Richardson observes in essays like the *Work of Art, Hölderlin and the Essence of Poetry,* also Being is essentially negatived and comes-to-presence as the dynamic tension between positivity and negativity, truth and non-truth: clearly it is *alētheia.* As in IM in these essays Being assumes the primacy over There-being. Here Being is a Dis-cord, i.e., the process of negatived truth, bestowing itself on There-being., 413-417.

[86]M. HEIDEGGER, *Holzwege* (Frankfurt am Main: Vittorio Klostermann 1950), 244.

[87]B.WELTE, "God in Heidegger's Thought," *Philosophy Today* 26 (1982) 90-91.

unconcealment as truth presents itself in Time.

6. BEING AND TIME

Being and Time, Time and Being - these are extremely elusive, and so is the relation between them. If Being is to be thought as fundamentally interconnected with time, then time must show itself when the Being of beings is questioned. Throughout his philosophizing, Heidegger stresses the point that Being is not a thing, not itself a being, so one cannot even say, 'Being is...' And what about time? Time is not a thing either. We say, everything has its time, but this is not some other thing that it has. Then how can time be primordially intertwined with Being? For the popular mind, time is an indefinite series of "nows", where the "future" consists in the "nows" that have not yet come but some day will be, the "past" consists of the "nows" that once were but no longer are, and the "present" is the "now" which at the moment "is." But obviously it does not answer the question of precisely what time is. Is time, and does time have a place? Heidegger argues that even the traditional characterization of Being as temporal, timeless, or supertemporal is a particular metaphysical interpretation which treats time as another thing we encounter like an object.[88]

As to the relation between time and Being, Heidegger fixes on the notion of presence. He claims that 'from the dawn of Western or European thinking until today, Being means the same as presencing.'[89] Presence speaks of the present, so that Being is determined by time and time is determined by Being. Thus, as Being-present, Being is determined by time. But in how far and in what respect is Being determined by time? Why, in what way, and from what is it that time resounds in Being? 'Being and Time'

---

[88]GA 24.434.

[89]TB. 2.

says Heidegger, determine each other reciprocally, but in such a manner that neither can the former, Being, be addressed as something temporal, nor can the latter, time, be addressed as a being.'[90] As Macquarrie suggests, we try to understand time not as a thing, but as the medium, as we might say, in which things arise and pass away. In a sense, time itself passes away, but in another sense only the things in time pass away, while time remains as time.[91]

In BT Heidegger interprets that being which is distinguished by the understanding of Being: man as Dasein in terms of temporality. From the temporality and historicity of Dasein who understands Being, Heidegger wanted to bring up for thought the time which belongs to the sense of Being.[92] As drive-towards-Being, There-being is constantly coming to Being, i.e., to its self. This coming of There-being to its self, and therefore of Being to There-being, is There-being's "coming," i.e., its future (*Zukunft*). By reason of this coming (future), There-being comes to its self-but to a self that already is as having-been-thrown. The self that already is as-having-been is There-being's past (*Gewesenheit*). By reason of the authentic comprehension of its own transcendence, There-being renders possible the encounter with other beings as beings. It renders them present. This rendering-present of beings is the authentic present (*Gegenwart*) of There-being. So it is that, in the structure of There-beings' transcendence, existence consists

---

[90]TB. 3.

[91]J. MACQUARRIE, *Heidegger and Christianity* (London: SCM Press 1994), 97. Hereafter J. MACQUARRIE, *Heidegger and Christianity*.

[92]Pöggeler asserts that in the unpublished Division Three of BT, Heidegger was going to determine the time-character of the transcendental horizon in which Being unfolds into its multiple meanings. But the attempt failed, and the third Division went unpublished. The investigation did not reach its goal. See O. PÖGGELER, "Metaphysics and the Topology of Being in Heidegger." 177.

in the coming (future) of Being to a self that already is (past), rendering manifest the Being of beings with which it is concerned (present).[93] "...This unified phenomenon we call temporality..."[94]

Temporality and historicity are introduced in BT as decisive 'existential' of the Being-comprehending Dasein and thereby as occasion for the problem as to how time itself belongs to the comprehended Being, to the meaning of Being.[95] Now the question arises as to how temporality should be considered in relation to this Being. It seems that temporality does not belong to Being as such, but to the finite realization of Being in beings. When Being is attended to only in the pre-ontological (hence pre-predicative) steps of its manifestations to Dasein, it is always found in the horizon of time. Thus viewed temporality is the very manner of becoming of finite beings. As Vadakethala points out temporality has nothing to do with the subsisting Being which is completely outside the realm of becoming.[96] From the temporal nature of Being Heidegger comes to the historical notion of Being. It is the temporal nature of Being, which allows itself to be

---

[93]BT. 370ff.

[94]BT. 374.

[95]The emphasis on time and temporality marks a breach with much of the philosophical tradition. Macquarrie clearly puts forward the problematic it involves when he says that in the philosophical tradition, the self was usually thought to be a substance. To deny substantiality to the self and to interpret it as a temporal structure would seem to rob it of claims to permanence. What is even more startling in Heidegger's philosophy is that Being as such appears to be temporal, and such a view is almost inevitable if one takes Dasein as the clue to the meaning of Being, and holds that Dasein through and through temporal. ... But just how the temporality of Dasein relates to, let us say, the time of physics is left unclear by Heidegger. After all, he did decide that there is no direct path from the analysis of Dasein's being to the general question of Being. J. MACQUARRIE, *Heidegger and Christianity*, 39.

[96]F.J. VADAKETHALA, *Discovery of Being*, 110.

temporalized and to be historicised, that makes history possible. The Being revealed with the horizon of temporality is thus historical Being. And Being is revealed historically because man could receive Being only historically.

The meaning of Being, in the phase after BT, seems to be determined anew as the Truth of Being. Historicity and history are to be experienced as belonging to this Truth. When Heidegger speaks of Being and Time, as Pöggeler suggests, time does not mean something which stands alongside Being, which perhaps must be superseded if Being itself is to be expressed. Being and Time are rather so intertwined that one can be understood on the basis of the other. Time is rather a primordial movement which Being itself, releases beings from out of itself.[97]

Karl Löwith has argued that Heidegger's expressions in his later writings often gravitate away from time and the historical and towards something like eternity. He singles out in particular Heidegger's references to the "abiding" (*das Bleibende*) and to "eternity" in *Erläuterungen zu Hölderlins Dichtung* and to the "indestructible" in the "Postscript" to *What is Metaphysics?*, all of which are for Heidegger's descriptions of Being.[98] In this "permanent" origin of beings, Heidegger seems to say, man finds rest from the succession of ages, from the sequence of epochal transformations. To the extent that Heidegger does move in this direction, his thought does indeed come closer to mysticism and begin to lose its "secular" character. For this attempt to escape from the fluctuations of time and to find "rest" in something "permanent" is distinctly typical of the mystics. Further, Heidegger claims that the "It" (*das Es*) which "gives" (*Es gibt*) is

---

[97]O. PÖGGELER, "Being as Appropriation," in *Heidegger and Modern Philosophy*, ed., M. MURRAY (New Haven: Yale University Press 1978), 85.

[98]K. LÖWITH, *Heidegger: Denker in dürftiger Zeit* (Göttingen: Vandenhoeck und Ruprecht 1965), 33-36. Hereafter K. LÖWITH, *Denker in dürftiger Zeit.*

not time - for time is also "given" (*Es gibt Zeit*) - but the Event (*Ereignis*) itself. The Event, therefore to the extent that it is the source of time, is itself beyond time.[99] Again this event of Being which is beyond temporality happens in the context of the world.

7. BEING AND WORLD

In keeping with the basic dictate of phenomenology, Heidegger tries to gain a "natural concept of the world." The human being is a worldly being; human existence is Being-in-the-world. The term *world* designates an essential dimension of human existence; it is an existential and not merely ontic concept. The expression "Being-in-the-world" describes the way of "to-be" of a human being; man and the world are equally original from the phenomenological standpoint.[100] The structure of Being-in-the-world belongs to the definition of the Being of the human being. The world, then, is not a thing, not even a conglomeration of all things, but the basic way of human existence,[101] a network of meanings. The term *world* in the final analysis, is understood not as the manifestation of a divine creative presence (as, for instance, in Augustine) but as a "network of meanings," as a "system of total meaningfulness."[102]

Pöggeler clarifies the point further when he says: As Being-in-the-world, life must be understood within itself; philosophizing must presuppose no Transcendence which would explain whatever is, especially once one suspects that the Transcendent, the God of philosophy, is not the God of faith. Life in its facticity always understands itself in some way; meaning belongs to it. Meaning in this search for Being, is no longer thought in terms of a schema where unchangeable eternity, residing statically in itself,

---

[99]TB. 40-42.

[100]G. KOVACS, *The Question of God*, 224.

[101]GA 24. 422-423.

[102]BT. Section 14.

is placed over against temporality as the changeable. Rather, meaning is to be conceived as the very movement of factical life which, in its originality, is history.[103]

The tendency in Western thought has been to think of the world as a collection of things set over against us, and of Dasein itself, not indeed as another thing, but as another entity 'present-at-hand.' But Dasein is not just another item in the world, not even another rather special item, to be designated by the word 'person.' Dasein is the 'being-in-the world.' The multitude of things contained in the world is seen and understood in the context of world, within which they are connected in a network with each other and with Dasein.[104] Things are seen by Dasein as ready-to-hand, available for use by Dasein in its dealings with world. Thus the things of the world become for Dasein equipment (*Zeug*) for living. For everyday existence, the world is a world of work. In the first instance, things are seen not as objects but are items of equipment for Dasein's tasks. The example of a thing which Heidegger chooses to illustrate his theory is a hammer. We see it as a piece of equipment for hammering, in human activities like building and furnishing. Things are ready-to-hand (*zuhanden*) in the sense that we incorporate them into our activities. Something which merely confronts us as an object is said to be present-at-hand (*vorhanden*), but increasingly we bring even such things into the realm of equipment. Hence, confining his analysis to 'everyday' existence Heidegger speaks of the world as a workshop.[105] This way of looking at the world in BT seems to be very much utilitarian.

In this context the following observations put forward by Kovacs seem to hold true. The human being is destined to the

---

[103]O. PÖGGELER, "Metaphysics and the Topology of Being." 175.

[104]BT. 95ff.

[105]BT. 100.

world and to itself; it transcends to the world as its ultimate meaning and not to God as the creator of its life and of the world. There are some vital questions regarding this phenomenological notion of the World that ought to be raised. Can the notion of God as the ultimate meaning and goal of all beings be reconciled with the concept of the world as the ultimate meaning, as the system of total meaningfulness? Can there be a meaning, at least to the human being, that is higher and more comprehensive than the world as described in Heidegger's phenomenology? Should the affirmation of the this-worldly destiny of human existence be viewed as the final, definitive resolution and identification of the ultimate meaning and destiny of the human person?[106] Heidegger's responses to many of these questions are neither complete nor decisive. The this-worldliness of human existence, the finiteness of the human being, the worldness of the world, and the existentiality of existence are endemic to his phenomenological methodology.

Heidegger also claims, however, and quite persistently, that his existential, phenomenological analysis does not decide the ultimate destiny and value of the human being; his definition of human existence, for instance, neither affirms nor denies the relationship of human reality to God.[107] The view on this relationship does have a bearing on the interpretation of human destiny as well as on the question of meaning. Heidegger's way of asking the question of ultimate meaning, therefore, does not seem to be truly comprehensive at least regarding the content of ultimate meaning beyond time or in timelessness.

However, it is interesting to note that Heidegger's essay called The Thing (*Das Ding*) paves the way for new understanding of the world which gets away from the somewhat exploitative

---

[106]G. KOVACS, *The Question of God*, 224.
[107]GA 24.316.

attitude to the world expounded in BT. The new view is expressed in terms of the 'fourfold' (*das Geviert*). A thing is not only more than an object, it is more even than a human product or an item in the human equipment for bringing the earth under control. A thing is now granted the possibility of having a beauty and dignity of its own. The fourfold means that everything has a fourfold reference, or has four dimensions of being which together constitute the meaning of that thing. The four dimensions are: earth and sky, mortals and gods. To illustrate his new understanding of the nature of a thing, Heidegger chooses the example of a wine-jug to elucidate what he means by the fourfold. The jug made of clay refers to earth. It refers to the sky from which has come both sunshine and rain to swell and ripen the grapes used for making wine. It refers to the humans, the skilled craftsman who has given to it the form. As Macquarrie suggests the word 'gods' does stand for what might be called a 'divine factor' in all reality, something holy in which every thing participates.[108]

According to Caputo the thought of Being (*Seinsdenken*) for which Heidegger appeals in all his later writings is a thinking which is meant to summon up a turn (*Kehre*) in this movement of Being. The fourfold is the new world, a world of unconcealment (*alētheia*) rather than of the technological darkening of the world (*Gestell*). The world of the fourfold means in particular the coming to pass of these four: the earth, the sky, the mortals and the gods. This is to be considered as a major step ahead from his early understanding of a secular and utilitarian considerations of the world.[109] The Being-in-the-world who is part and parcel of the world comprehends the Being and the Being manifests itself in the comprehension of the thinking of being.

---

[108]J. MACQUARRIE, *Heidegger and Christianity*, 65.

[109]J.D. CAPUTO, "Heidegger's God and the Lord of History," *New Scholasticism* 47/4 (1983) 450.

8. BEING AND THINKING

The question about the sense of Being remains through Heidegger's entire work an indefatigable effort to think the Being-process. The thought that thinks the Being-process is the thought which interrogates the foundations of metaphysics. Heidegger calls it foundational thought. As opposed to the tendency to dominate the objects of thought, the attitude of foundational thinking will be simply to let beings be, hence render them free unto themselves. More positively, foundational thinking tries to meditate Being as the process of truth, i.e., the coming-to-pass of the lighting-process in beings. It is brought-to-pass by the nature of man conceived as ek-sistence, of an ecstatic open-ness unto the lighting-process of *alētheia*. Ek-sistence thus understood may be called the There (*Da*) of Being, because it is that domain among beings where the lighting-process takes place.

In *Letter on Humanism* Heidegger tells us that thinking is the thinking of Being. Thinking is of Being in as much as thinking, coming from Being, belongs to Being. At the same time, thinking is of Being insofar as thinking, belonging to Being, listens to Being.[110] This implies the primacy of Being over the beings and over Dasein, and this is made explicit: Being itself 'presides over thinking and hence over the essence of humanity, and that means over its relation to Being.'[111] As we have already noted what Heidegger is saying here is also an attack on the metaphysics of subjectivity, the belief that the human mind is the measure of everything. To get away from any hint of subjectivism, Heidegger writes in this essay 'ek-sistence' rather than 'existence,' stressing that it denotes a 'standing out into the truth of Being.'

According to Heidegger Being needs thinking for its self-manifestation. For, all along, Being for Heidegger is the Being

---

[110]LH. 196.

[111]LH. 196-7.

which reveals itself in and through human thought. Hence in his view, Being needs and uses man.[112] Heidegger is intent on thinking the meaning of Western history, a history which he conceives not in the ordinary sense as a sequence of happenings, but as a history of truth and untruth, as a play of light and darkness, of successive clearings in which the various "worlds" of Western historical man are rendered open and manifest.

For Heidegger the essence of philosophy is questioning and thinking of Being, of the ontological difference. Philosophy is thinking as the thinking of Being. The main thrust of Heidegger's thought is the unfolding of the full dynamics of questioning; it is the search for and not the possession of answers. What Heidegger advocates is meditative thinking rather than the representational thinking of the Western tradition. Representational thinking is depicted by Heidegger in the "Postscript" to *What is Metaphysics* as an attempt to calculate and count up beings, to reckon upon the possible uses to which they can be put.[113] Essential thinking, on the other hand, "spends itself" (*verschwendet*)[114] not on beings but on the truth of Being itself. The essential thinker puts aside as "images" and "representations" to engage in a non-representational thinking on Being which leaves metaphysical reason behind.

Heidegger discusses the attitude and posture of "essential thinking" vis-à-vis Being. By means of thinking Dasein 'surrenders' its being (*wesen*) to the simple necessity - a necessity without force - which man is under to think the truth of Being. This surrender is a "sacrifice" (*Opfer*) and in this sacrifice man must give himself up to the truth of Being. This sacrifice is a

---

[112]J.D. CAPUTO, *Heidegger and Aquinas: An Essay on Overcoming Metaphysics* (New York: Fordham University Press 1982), 281-282. Hereafter J.D. CAPUTO, *Heidegger and Aquinas*.

[113]WM. 356-357.

[114]WM. 358.

"thanking" (*Danken*) of Being by Dasein for the "grace" (*Gunst*) and favour (*Huld*)[115] which Being has bestowed upon Dasein by giving itself up to Dasein as a matter for thinking. "The sacrifice is a detaching from beings on the road to the preservation of the favour of Being.[116] Since Being does not submit to human control then, if Dasein has succeeded in thinking, it is because Being has "given" itself to Dasein, has bestowed in effect a grace upon it. The appropriate response of Dasein to a grace is thanking, and its thanking consists in not squandering this gift but in actually engaging in "thinking." Now Heidegger asks rhetorically how else "thinking" is possible for Dasein unless the grace of Being preserves for man, through man's open relationship to Being itself, the nobility of poverty in which the freedom of the sacrifice conceals the treasure of its essence?[117]

Heidegger writes of both thanksgiving and sacrifice in the human being's relation to Being. The thinking of Being is also a thanking. 'Original thanking,' he writes, 'is the echo of Being's favour, wherein it clears a space for itself and causes the unique occurrence: that the beings are.' And 'sacrifice is taking farewell of all the beings on the way to the maintenance of the favour of Being.'This language about taking farewell in Macquarrie's opinion is, directly reminiscent of the mystic, Meister Eckhart.[118]

As time passed, Heidegger preferred to call himself a thinker, rather than a philosopher. He also came to believe that philosophical thinking has more in common with the thinking of the poet than it has with the thinking of the scientist. For Heidegger firmly believed that the poet, just as much as the scientist or the philosopher, is concerned to speak the truth, that

---

[115]WM 358.

[116]WM. 359. *Das Opfer ist der Abschied vom Seienden auf dem Ganz zur Wahrung der Gunst des Seins.*

[117]WM. 358-359.

[118]J. MACQUARRIE, *Heidegger and Christianity*, 53.

is to say, to uncover Being. Therefore in the postscript to WM, he says: 'The thinker utters Being. The poet names the holy.'[119] Thus being makes its way towards Being - a movement from being to Being.

9. BEINGS TO BEING

In the decades following the publication of BT, Heidegger's *Seinsfrage* shifted its focus. The epoch of BT and of what is often referred to as the earlier Heidegger is followed by the epoch of later Heidegger. The problem of the ontological difference and the emphasis on the primacy of Being become more and more pronounced with the unfolding of Heidegger's way of thinking, especially in the later works. This change of emphasis from "There" to Being could be regarded as the consequence of the meditative experiencing of the coming about of the ontological difference and of the deepening of the question of Being on Heidegger's way. Heidegger referred to this change as a 'turn' and a 'completion' of the questioning that had begun with BT. If BT had analyzed 'man in relation to Being,' Heidegger characterized the 'turn' with an inversion in terms, as an approach to 'Being and its truth in relation to man.'[120] Heidegger himself tended to minimize the extent of 'turning' that had taken place in his thought and interpreted the new ideas as developments of what was already implicit in the early writings. Nevertheless, he

---

[119]WM. 360.

[120]Heidegger's acceptance and interpretation of the distinction between the earlier phase and the later development of his thought can be found in his "*Vorwort*" (Letter to Richardson, 1962), in W.J. RICHARDSON, *Phenomenology to Thought,* xx-xxii. Insightful presentations and interpretations of this change in Heidegger's development can be found in: O. PÖGGELER, ed., *Heidegger: Perspektiven zur Deutung seines Werks* (pp.11-54); A. De WAELHENS, "Reflections on Heidegger's Development: Apropos of a Recent Book," *International Philosophical Quarterly*, 5 (1965) 475-502.

plainly admitted that the path which he had first tried to follow had broken off, and this fact is amply attested by the unfinished state of BT.

Obviously there are quite important differences between what we find in his early thinking, as expressed most comprehensively in BT, and what we find him saying at later stages in his thought. Perhaps the most obvious is that there is a shift of focus from the Dasein known in human existence to Being in the most universal sense. A second point is that the phenomenological method used in the analysis of Dasein gives way to an appreciation of the language of the poet in interpreting the meaning of Being and Dasein's relation to Being. There is a change too in the character of the thinking. Whereas the early thinking of the philosopher is investigative and active, the later thinking becomes more and more meditative. The change in the concept of world is also notable. In the early thinking, the world is an instrumental system, and things lie ready-to-hand for the use of the Dasein in everyday concerns. But in the later writings, the world is no longer primarily a workshop but has a dignity in its own right, so to speak. Things are not just 'equipment' but are constituted by the 'fourfold' of earth and sky, mortals and gods.

Now, what concerns us most is the apparent difference between Heidegger's original view on the relationship between Being and man and the one which we find in his later works. In BT, Heidegger suggests that man plays the leading role in this relationship, whereas in the later works it is clearly Being itself which holds the primacy in its own self-disclosure. Furthermore, Being itself, which in BT was described as the total-meaningfulness of the world which is projected by Dasein, is later described as that which gives itself to man, determines him, and even dominates him as its *Da* among beings. In BT, Heidegger had stated that only as long as Dasein is, "is there" Being, and only as long as Dasein is, "is there" truth; in the later works, however, it is made clear that the *Da* itself is thrown, and that it is thrown by Being itself.

Finally, whereas in BT the coming-to-pass of Being is said to take place in and through Dasein's care, in the later works we find that it is Dasein's thinking which is to occupy this important place. It points to the fact that the stress of accent has been shifted from Dasein to *Sein* as such. Thus the post BT publications as well as his reflections on Hölderlin, with their difference of procedure and shift of emphasis could be considered as a second phase of Heidegger's thinking or the 'change in the later Heidegger as authors would like to put it. Vadakethala adds when he says that Heidegger proposed the analysis of Dasein in BT only as a preparatory analysis to clear the way to the understanding of Being. Being was the goal. The later writings are an attempt to arrive at this goal.[121]

In BT, the question concerning the meaning of Being is raised but not explicitly treated as such. Heidegger states there that it is the function of ontology to concern itself with this question, whereas *fundamental ontology* must focus on the mode of Being of the one who asks the question, i.e., the mode of Being of Dasein. Because the book was published in an incomplete form, the Being question itself was never explicitly discussed in BT. This means that even though the purpose of BT was primarily to pose the Being-question, in the event it proved to be principally an analysis of the Being-comprehension of There-being.

As Macquarrie suggests, in his *Introduction to Metaphysics*

---

[121]F.J. VADAKETHALA, *Discovery of Being*, 11-13. He observes three phases in Heidegger's thinking. The first, Dasein as presented in BT; *Sein* as presented in later writings; and the third, the ontological difference as presented in still later writings. Nevertheless already in *Kant and the Problem of Metaphysics* (1929) one notices reference to the question of ontological difference. Hence different phases in Heidegger's thinking are not strictly chronological rather they are stages of thinking itself.

Heidegger reintroduces the question of Being.[122] Although IM discusses the essential elements of BT (structure of There-being: transcendence, finitude, temporality), they are dealt with a different accent, for now Being maintains the primacy over There. To think Being thus conceived in Richardson's view, is to bring-to-pass what the early Greeks meant by *Logos* and *noein*.[123] The There is opposed to Being, but is not separated from it in subject-object fashion. Its function is to gather into concentration the overwhelming power of Being and thus contain its dynamic advance in such a way as to force it into the disclosure through which the non-concealment (truth) of beings comes-to-pass.

The *Letter on Humanism* crystallizes Heidegger's conception of the relation between Being and man. Man is a "mortal", cast forth by the "throw" of Being. His true worth lies in guarding and watching over Being as its shepherd (*der Hirt des Seins*).[124] Being gives itself (*es gibt das Sein*), it brings itself into openness. Being, as bringing itself into openness, is disclosure, the truth, the *Da*. To guard the truth of Being, to let this truth abide in him, i.e., to be the *Da* of Being, is the mission of man which makes man man. The essence of man lies in his ek-sistence, i.e., an ecstatic relationship to Being; Being in turn is not only in relationship with There-being but *is* this relation inasmuch as it imparts itself to man. It seems to say that to speak of man's essence is to speak of the Being of beings.[125] After examining the radical reinterpretation which Heidegger has made in LH of such basic concepts in BT as thrownness, resoluteness, truth, and the term 'Da-Sein' itself, Löwith concludes that Heidegger has subscribed not to one but two grounds in the course of his work. The first is Dasein, on the basis of which 'there is' (*es gibt*) Being (*Being and Time*). The

---

[122] J. MACQUARRIE, *Heidegger and Christianity*, 54.

[123] W.J. RICHARDSON, *Phenomenology to Thought*, 297.

[124] LH. 234.

[125] LH. 228.

second is Being itself, on the basis of which there is (*es gibt*) Dasein (the later writings).[126]

We have seen often enough that Being for Heidegger is fundamentally a lighting process by which beings are lit up as what they "are." This supposes, however, that among them there is a domain where this lighting up process takes place. This is the There of Being. Now in this coming-to-pass, Being maintains the initiative. For the There is "thrown," and it is Being that does the "throwing."[127]It is Being that gives itself unto the There with the There and, imparting itself thus, continues to dominate its There. The coming-to-pass takes place for the sake of Being, and the There, thus brought-to-pass, sustains the process, guards it, stands watch over it so that in the light of Being beings may appear non-concealed. In other words only insofar as the lighting-process comes-to-pass in a There does it come-to-pass at all. This coming-to-pass, however, does not depend on the There but on the spontaneity of Being which thus e-mits itself among beings.That which is thus e-mitted by Being (*ge-schickte)* is what makes history possible (*Geschichte*). History is that process by which being comes-to-pass by self-mittence.

In BT Heidegger focused mainly on the relation between Being

---

[126]K. LÖWITH, *Denker in düftiger Zeit*, 15ff. Like Löwith, Versènyi sees an abrupt reversal between the standpoint of BT and the point of view defended in LH. For Heidegger's position in BT "is so Dasein-related, man-bound and relativistic" that it is "surprising" to see Heidegger a decade or so later take "strong exception to what he considers to be humanistic, subjectivistic theories of truth." Although the later writings shift all the emphasis to Being, still he finds in both BT and the later work a bond between Being and Dasein which is such that "there is" Being only while "there is" man, and there is man only so long as there is Being. Between Being and Dasein there exists a "correspondence"- Being "addresses" man and man "responds"- a living interchange. L. VERSENYI, *Heidegger, Being and Truth* (New Haven: Yale University Press 1965), 52f.

[127]LH. 234.

and one particular kind of being, namely, Dasein; and it is often suggested there that the primacy in this interaction belongs to Dasein. Heidegger described this relationship in terms of transcendence and in view of the fact that Being concretely always has the structure of world. The term world is often used instead of Being. Now, that which is transcended (beings) belongs as essentially to the process of transcendence as that whereunto the transcendence is oriented (world, Being), so that already in BT Being appears as the Being of beings.[128]

In the later works, Heidegger begins to place the stress on the coming-to-pass of the truth. He then maintains that Being is the Being of beings, in that Being is that by which beings are, and Being is drawn toward beings in which it must emerge in order that it be itself.[129] Although it is said that Being is the Being of beings in the earlier works also the correlation analyzed by thinking is that between thought and Being. Whereas in the later works Heidegger began to focus more directly on the ontological difference as such, so that the correlation analyzed is now one between thought and the Being *of* beings, expressing that Being means the Being of beings and beings mean what they are in terms of their Being.[130] In the postscript to WM, (1943) Heidegger suggests that Being can be and even must be thought by itself, independent of the beings. "It pertains to the truth of Being that Being indeed comes-to-presence without beings..." even though beings are never without Being.[131] Yet in most other later works, we constantly find the claim that Being is nothing but the Being of beings, and in 1949 Heidegger corrected the statement of the postscript to read "Being never comes-to-

---

[128]J.J. KOCKELMANS, *Truth of Being*, 56.

[129]M. HEIDEGGER, *What Is Called Thinking?*, tr., F.D. WIECK & J.G. GRAY (New York: Harper & Row 1968), 4-5.

[130]WT. 226-227.

[131]WM. 354.

presence without beings."

Richardson's assertions seem to be rather a good attempt at answering the seemingly contradicting statements of Heidegger. He clarifies the change in readings as follows: The first text insists on the primacy of the Being-process in the emergence of beings, therefore names the ontological difference, but not as such. Whereas the second text insists on the fact that although Being must be thought for itself in order to comprehend the ontological difference, it can never *be* by itself. Hence, if it is true to say that beings cannot be without Being, the reverse is also true: Being cannot be without beings. This names the ontological difference *as such.* Briefly the formula of 1943 emphasizes the primacy of Being and implies the ontological difference but does not name it as such. The formula of 1949 names it *as such.*[132]

If we approach Being itself from the perspective of beings, then Being appears as somehow encompassing them, just as a domain of openness encompasses what is found within it. This domain is not space, however, but rather that openness out of which even space and time themselves come-to-presence. Furthermore, Being must not be understood as something static; it is a process, the clearing process by which beings are lighted up. When man enters into a comportment with beings, this light is neither man nor the beings; rather, it is the "between" that is between them as that which makes the encounter take place.[133] Later Heidegger adds that the open which lets be is the same as that which originally was called Non-concealment and which, at the beginning of Western thought, was designated as *a-létheia*, truth.[134] Finally Being is not an existing quality of beings, nor can Being be conceived and established objectively. Yet it is this non-

---

[132]W.J. RICHARDSON, *Phenomenology to Thought*, 562-565.

[133]J.J. KOCKELMANS, *Truth of Being*, 57.

[134]M. HEIDEGGER, *The End of Philosophy*, tr., J. STAMBAUGH (New York: Harper and Row 1973), 91.

entity that gives every entity the warrant (*Gewahr*) to be.[135]

In his essay *Discourse on Thinking (Gelassenheit)* Heidegger uses the term *the open* to refer to Being. Being is understood there as the open domain, the broad region of expanse (*Gegnet*) which is completely open and free, wherein all beings may while, and which gathers together every being to itself and all beings unto one another in the "process" which whiles and rests within itself.[136] Being is here understood primarily as the gathering process, *logos*; yet it is also called *die Gegnet*, "the expanse," because it is the open domain which, in gathering beings together, opens itself up in such a way as to establish and maintain the open which permits every being to emerge and rest within itself. [137] Again Heidegger sees Being as a sphere because it is Parmenides' "One" (Ἕν), which he understands, as the process of coming-to-presence in all beings which is everywhere the same, hence a circle whose centre is in every being, lighting it up from the inside as what it is. The spherical element of the One and the One itself have the character of a lighting-up process that reveals (beings). Within this process (beings) can come to presence.[138] "Being, indeed - what is Being?" Answering the question he writes, "It is It itself." He continues thus:

> "Being" - that is not God and not a cosmic ground. Being is farther than all beings and is yet nearer to man than every being, be it a rock, a beast, a work of art, a machine, be it an angel or God. Being is the nearest. Yet the near remains farthest from man.[139]

As far as man is concerned, Being itself is farther away than all beings; yet it is also nearer to him than every being. Being is

---

[135]WM. 353.

[136]M. HEIDEGGER, *Discourse on Thinking*, tr., J.M. ANDERSON & E.H. FREUND (New York: Harper & Row 1966), 67-68.

[137]DT. 68.

[138]PLT. 123.

[139]LH. 234.

nearest to man, insofar as it is that which lets him be what he is and makes it possible for him to enter into comportment with beings. It is farthest from man, insofar as it is itself not a being and insofar as man is structured in such a way that he can deal directly only with being. Thus, Being is the mystery. From the point of view of beings, Being encompasses them all, just as a domain of open-ness encompasses what is found within it. This domain is not, of course, "space" but rather that dimension out of which even space and time themselves come-to-presence. Being is the domain of open-ness, because it is the lighting-process by which beings are lighted-up. Being, at once "in" man as "in" every entity and beyond his grasp, has a view of man, manifests an intention, imprints a direction from which man cannot extricate himself.

Being ap-propriates to man his own proper nature, which consists in the fact that he is "released" (*Gelassenheit*) unto Being.[140] Thus it is that from the very beginning, i.e., according to the very origins of his nature, man belongs to Being as its own. Although Heidegger does not put it into these words, that in ap-propriating to man his nature, Being ap-propriates him to itself, and it is this double ap-propriation that we try to suggest by saying that man by nature is an attend-ant of Being, whose proper task is to attend to it.[141]

---

[140]In this context Harr observes the following: "Being "uses" man: *braucht ihn*. Man is not simply "maintained" ["*main-tenu*"], but literally *manipulated* by being because the relation that is termed *Brauch* translates or transposes Anaximander's *Chreon* (from *cheir*, the hand [*la main*]). Pointing to the belongingness of being and Being, he asks furhter, do we not find the transposed taking up again of a religious theme and its current expressions: "we are in the hand of God", or "God has need of man" (*brauchen* means both "to need" and "to use"), or again, "Man proposes, God disposes"? M. HARR, *Essence of Man*, 64.

[141]TB. 40-42. The German expression *es gibt*, literally, 'It gives,' is to be taken in the literal sense, 'It gives Being.' The 'It' that here 'gives' is Being

10. BEING AS EVENT

Being marks each being by coming-to-presence, insofar as it, by coming-to-presence, lets beings be present, unveils them, and brings them into the open. In the process of unveiling, there is a kind of granting at work which grants Being-present, while it lets-be-present that which is present, namely, beings. In this process we come again upon a granting, and thus upon an "it" which grants.[142] Heidegger remarks: We try to bring the it and its giving into view, and capitalize the 'It.'[143] Heidegger says clearly that Being is a gift of the 'It gives' and that Being belongs to giving. 'It gives Being' means that there is presence, that there is unconcealment, that Dasein is brought into the clearing. As Macquarrie claims, it is the 'It gives' that is more ultimate even than Being and seems to come close to what has ordinarily been understood as God. He continues: "In Christian theology, God is love. In Heidegger, 'It gives' is an act of giving or donation, and since he has told us that the 'It' which gives Being is Being itself, then the act of giving is also an act of self-giving, and so not different in any major respect from love."[144]

Heidegger asserts that the essence of Being itself is self-giving. The 'gives' names the essence of Being that is giving or granting its truth. The self-giving into the open, is Being itself.[145] Further, he says that Being is not the product of man or Dasein, and that 'man is not the lord of the beings, but the shepherd of Being.'[146] Thus he denies the idea that man is the master of the world or the measure of all things. Being is not something which is found

---

itself.

[142]TB. 36-38.

[143]TB. 5.

[144]J. MACQUARRIE, *Heidegger and Christianity*, 99.

[145]TB. 37-40.

[146]LH. 234. It reads: '*Der mensch ist nicht der Herr des Seienden. Der Mensch ist der Hirt des Seins.*'

outside the granting, as is the case with a common gift. In the granting, Being as Being-present becomes changed. As letting-be-present, it belongs to the unveiling itself, and as gift it remains contained in the granting.

Being as the unveiling of Being-present is granted by the mysterious It.[147] As regards Being, this giving is active, a letting-be-present, unconcealing, admitting into the open. It is the given only that is brought into the open, not the giving. Both the giving and the It that gives remain hidden. A giving that thus withholds itself is a sending (*Schicken*), and the gift of Being is a mittence (*Geschick*). It is in the essay on *Nietzsche's word "God is Dead,"*[148] Heidegger uses the word mittence which points to the event in which Being is disclosed, proceeding from its own initiative. For Heidegger it designates an event, hence a coming-to-pass, by which Being "sends" (*sich schickt*) itself unto man affirming its own primacy.

In his lecture on *Time and Being*, Heidegger seems to name the 'It gives' as *Ereignis* (event).[149] He says: What determines

---

[147]TB. 5-6.

[148]M. HEIDEGGER, "Nietzsches Wort 'Got ist tot'," in *Holzwege*, 193-247.

[149]For a helpful discussion of the connotations of *Ereignis* and the strands of thought behind Heidegger's use of the term, see Hofstadter's introduction to his volume of translations, PLT, xix-xxii. The thought of Being as event (as appropriation) prepares another language for Being which names Being in several ways. It is interesting to note the different names for Being listed by G. Kovacs from GA 65 which are quite descriptive: appropriation/event (*Ereignis*), fissure (*Zerklüftung*), strife (*Streit*), uniqueness (*Einzigkeit*), the "between" (*Zwischen*), abyss (*Abgrund)*, primal ground (*Urgrund*), push (*Stoss*), strangeness (*Befremdlichkeit*), Beying (*Seyn*), otherness (*Andersheit*), clearing, lighting process (*Lichtung)*, as finite (*endlich)*, life (*Leben*), inexhaustibility (*Unerschöpflichkeit*), necessity (*Notwendigkeit*), potentiality (*Möglichkeit*), as "recoiling upon itself"(*Zurückgebogenheit) as the nothing (als das Nichts), excess (Übermaß*). See G. KOVACS, "The leap (*der Sprung*) for Being in Heidegger's *Beiträge zur Philosophie (Vom Ereignis)*", *Man and World* 25(1992) 49.

both Time and Being in their own, that is, in their belonging together, we call *Ereignis*, the event of appropriation.[150] The event means Being as such, insofar as it is the process through which the ontological difference issues forth.[151] "That which brings about the event is the event itself, and nothing else besides."[152] In Heidegger's view, the truth of Being comes-to-pass in singular ways. Each self-manifestation of Being in beings has the character of an *Ereignis*, an appropriating event. In each concrete case, the appropriating event binds together Being and beings; it weaves Being, man, things, and world together into an articulated and textured whole. It is the appropriating event which in each case lets man come into what he most properly is; it is what first launches man into his essence.

According to Harr, Heidegger develops the difficult notion of *Ereignis* to think the tie between man and Being in a quite different way. Their joining would precisely not be a link in the

---

[150]TB. 19. The German verb *ereignen* usually means simply 'to happen.' while the noun *Ereignis* is simply 'event.' *Ereignis* includes the root *eigen* corresponding to the English word 'own.' The *Ereignis* then is not just any event, but the event of appropriation, when something is made someone's own. This appropriation would seem to be the other side of the act of giving. So we read: 'The gift of presence is the property of appropriating.' (TB. 22). So this ultimate source is not a substance but an event which 'gives.' Heidegger argues that the word *Ereignis* cannot be properly translated, since it is as basic to this thinking as was the Greek *Logos* or the Chinese *Tao* to theirs (ID. 36).The word appears to contain the root *eigen*, "own, proper," which can suggest man coming into his own, finding his self. Translators generally use neologisms based on "appropriate" or "own." Thus we have "appropriation" (J. Stambaugh), "disclosure of appropriation" and "enownment"(A. Hofstadter). The connotation of an event is captured in "e-vent" (W.J. Richardson). Both are present in "event of appropriation" or "appropriating event" (T. Kisiel and J. Kockelmans), and "propriative event" (D.F. Krell).

[151]ID. 65-66.

[152]WL. 127.

sense of a relation "with or without dialectic," in the sense of an *attachment* of two terms originally separated, but of an initial identitiy, a full and unlimited "belonging together," a "constellation," an "interlacing," or "intertwining" (*Verflechtung*).[153] *Ereignis* as an inter-belonging of Being and man, as "the belonging together of call and obedience" (*das Zusammengehören von Ruf und Gehör*),[154] would be more original than the relation considered in terms of its two moments. Only a "leap" of thinking, a leap that "leaps away from being,"[155] would allow us to understand this wholly reciprocal appropriation, allow us to see that Being *belongs as much to us* as we belong to it, since "Being itself belongs to us; for only in us can it essence, i.e., presence as being[156] "The leap is the abrupt entry into that realm from out of which man and Being have always already reached one another in their essence, because both are appropriated over to one another out of a reaching-towards (*Zureichung*). The entry into the realm of this transpropriation first attunes and determines (*stimmt und be-stimmt*) the experience of thinking."[157]

Being reveals itself in its meaning, or in its openness and truth, as the event of appropriation. The *Ereignis,* says Pöggeler, does not mean a certain occurrence or happening, but rather Dasein's complete self-realization in Being, and Being's appropriation (*zueignen*) to Dasein's authenticity.[158] *Ereignis* determines the meaning of Being itself. It is, as a *singulare tantum*, a key concept

---

[153]M. HARR, *Essence of Man*, 65.

[154]M. HEIDEGGER, *The Question of Being*, tr., W. KLUBACK & J.T. WILDE (London: Vision Press 1974), 77.

[155]ID. 32.

[156]ID. 33. *denn nur bei uns kann es als Sein wesen, d.h. an-wesen.*

[157]ID. 33.

[158]O. PÖGGELER, "Being as Appropriation", 102.

of thought like the Greek word *logos* or the Chinese word *tao.*[159] One of Heidegger's main thrusts of the word "*Ereignis,*" in Caputo's view is to emphasize that by it Being comes into its "own" (*eigen*), that the "thing" is released into its own most Being, and enters into man's own most essential nature. In the Event (*Ereignis*) man is appropriated (*ver-eignet*) or brought forth into his own most proper being, for his own most being is to provide the place of disclosure for the revelation of Being.[160]

Now Heidegger claims that the event of appropriation is without a "why?" which asks about grounds or reasons. "It remains," Heidegger says at the conclusion of his lecture on *Der Satz Vom Grund*, "just play: the highest and the most profound play. But this 'just' is everything, the one, the only."[161] Heidegger's stance above or behind ordinary ways of talking raises the suggestion that he is trying to renew aspects of the mystical traditions. Hence Zimmermann observes: In Dasein (self) as finite openness, there is a relation to Being comparable to the mystic's relation to God. This comparison has been contested on the ground that in Heidegger there seems to be no personal or loving element in the relation of the Dasein to Being or the Event or however one designates the Ultimate Reality in Heidegger's philosophy. But he adds that Heidegger's releasement is a gift or loving bestowal which allows us to think, thank and love. He refused to personalize the Event (*Ereignis*) because this would

---

[159]ID. 36.

[160]J.D. CAPUTO, *Mystical Elements*, 224.

[161]M. HEIDEGGER, *Der Satz vom Grund* (Pfullingen: Neske 1957), 188.Texts such as this give some weight to the opinion that Heidegger's thought is in some degree "mystical." For a survey of the mystical connotations in Heidegger's thought, see J.D. CAPUTO, *Mystical Elements*; R. SCHÜRMANN, "Heidegger and Meister Eckhart on Releasement," *Research in Phenomenology* 3 (1973) 95-119; C. PAX, "Companion Thinkers: Martin Heidegger and Saint John of the Cross," *Philosophy Today* 29 (1985) 230-244.

have the effect of treating it as a being.[162] "*Ereignis* appropriates man to the usage (*Brauch*) it makes of him."[163]

All schemas of freedom, activity, and power that were those of Being are transferred to *Ereignis*, in the face of which man keeps the same determinations of quasi-passivity and dependency. In the Event man comes to himself, discovers his identity and relationship with the Being. The thought of *Ereignis* therefore does not fundamentally change the definition of man's essence as resulting from his relation to Being, and from Being's relation to him. *Ereignis* is therefore the name given by way of anticipation to a possible 'identification' between man and Being, beyond metaphysics - to the groundless ground.

11. BEING: GROUND AND GOD

Heidegger in search of Being wants to get to its fundamental realm in many different ways. He uses Leibniz's principle: *Nihil est sine ratione (Nichts ist ohne Grund)* in tackling the problem of ground in *Der Satz vom Grund*.[164] It is a fundamental principle of all human thought, and Heidegger even suggests that it is the basic principle of all principles.[165] The principle states something about every being, namely that it has a ground. Typically Heidegger hopes to get beyond that to the source of every being-Being itself. The German term *Satz* does not only mean principle but also a leap[166] and that it is this leap which will turn the

---

[162]M. ZIMMERMAN, *Eclipse of the Self* (Ohio: University Press 1981), 248-249.

[163]WL. 130.

[164]The evaluation of the principle differs considerably from that of *Vom Wesen des Grundes* (1928). For an understanding of the concept of ground in *Vom Wesen des Grundes*, see K. DE. KUYER , "The Problem of Ground in the Philosophy of Martin Heidegger," *The Thomist* 47 (1983) 100ff.

[165]SG. 48; 84.

[166]SG. 96.

principle from a principle of being into a principle of Being. In the statement 'nothing "is" without ground,' there is a harmony of "is" and "ground." "Is" naturally indicates in some way the Being of that which is (*Sein des Seiendes*).[167] This shows that the principle is not merely a principle of being but also a principle of Being. Hence Heidegger says: "To Being pertains such a thing as ground. Being is of the nature of ground. Being is in itself as grounding. Heidegger goes on to show that Being and ground pertain to each other and concludes "Being and ground: the same."[168]

As outlined in ID, Heidegger always regards identity as a co-belonging which simultaneously holds two things together and apart. We have already seen that in the inaugural lecture and its later additions there has taken place the massive shift in primacy from Dasein to Being. It is not Dasein who invents the concept of Being, but rather Being that thinks in Dasein. In the Postscript it is said that 'Being may be without the beings, but never the beings without Being.'[169] This suggests that Being is somehow independent of beings. Dasein in so far as it is ek-sisting to the Being of being is really the ground of metaphysics. But in so far as it goes out it acknowledges its own incompleteness and incapability of grounding metaphysics absolutely. As Vadakethala argues the ek-sisting ground by the very fact that it is ek-sisting calls for another ground free of this ek-sisting character which is

---

[167]K. DE. KUYER, " The Problem of Ground in the Philosophy of Martin Heidegger", 109.

[168]SG. 90.

[169]However, as Richardson points out, the Postscript reprinted along with the Introduction in 1949, there had taken place an unannounced reversal of the statement about the asymmetrical relation between Being and beings. Now it was said that Being needs the beings just as the beings need Being. See W.J. RICHARDSON, *Phenomenology to Thought,* 562-565.

the subsisting ground of metaphysics.[170]

While Being "is" dispensationally identical with ground, in itself it is groundless. But why is Being groundless? Because every form of grounding, even self-grounding would reduce Being to some kind of being.[171] In Heidegger's usage, the true "ground" of human existence is Being itself. Thus man is uprooted if he is pre-occupied with beings, to the neglect of Being. But by the "leap" of thought, we will arrive at a ground, a new rootedness which is Being itself. In so doing we will attain a new way of thinking which is not confined to "giving reasons." SG tells us that Being and ground are to be understood in terms of one another: Being and ground belong together. Ground receives its essence out of its belonging together with Being as Being. Conversely, Being as Being holds sway out of the essence of ground.[172]

In saying that Being and ground belong together, Heidegger does not mean to say that Being "has" a ground but that Being "is" ground. Being does not have a ground but itself serves as the ground for every being. Being is itself ground-less (*grund-los*),[173] i.e., not grounded by anything else. Otherwise we would face an infinite regress, always looking for the ground of the ground of beings, etc. Inasmuch as Being is a ground without ground, Heidegger calls it the "abyss" (*Abgrund*) that which lacks a basis and ground. Being then is both ground and abyss. In the traditional key, the principle of ground is a proposition which belongs to the realm of "representational" thought. But Heidegger's new intonation demands a more simple kind of "thinking" of Being itself, of "Being as Being, i.e., as Ground."[174]

---

[170]F.J. VADAKETHALA, *Discovery of Being*, 120.

[171]SG. 185.

[172]SG. 92.

[173]SG. 93.

[174]SG. 96.

Being and ground belong together. But instead of being thought together with Being, ground becomes a determination of the being, and the search is undertaken to find the ground of one being in another. What Heidegger calls for is a non-representational experience of Being, and not, as in BT, for a new conceptual determination of it. In Caputo's view like the mystic Eckhart Heidegger thinks his way through not to a first cause or ground but to an "abyss" (*Abgrund*) or "nothingness." Heidegger's "way" to Being is not the way of "discursive" reason, but the way of meditative stillness and total openness to that which is wholly other than beings, to the "simply transcendent."[175] To think Being adequately we must let Being be, let it lie forth (*vorliegenlassen,legein*) and emerge of itself (*physis*). Being is ground - i.e., Being and ground: the same - there is no need to supply a ground for it. As that for which every kind of ground and explanation remains inappropriate,[176] Being is thus the "abyss" (*Ab-grund*).

In Lotz's view the critical flaw in Heidegger's account of Being as ground is his notion of Being as a groundless ground, as a veritable abyss (*Ab-grund*) Being is indeed an abyss, Lotz holds, only if one means thereby that Being is a ground which rejects any ground other than itself, that is, if Being is self-grounding (*ens a se*). For Being in contradistinction to beings is the absolute fullness of all possible ways to be, and so there can be nothing other than itself upon which it is founded. If there were, Being would not be Being.[177]

Now the experiencing of the meaning of Being, the comprehension of Being, is somehow parallelled by the experiencing of God's presence or absence. However, the

---

[175]J.D. CAPUTO, *Mystical Elements*, 29.

[176]SG. 185.

[177]J.B. LOTZ, *Martin Heidegger und Thomas von Aquin: Mensch, Zeit, Sein* (Freiburg: Herder 1975), 78-82.

ultimate problem will be to find the right way to interpret the Heideggerian notion of Being and to grasp its unclarified relationship to the notion of God. The earlier phenomenological treaties seemed to have been composed under an epoché which neutralized all religious phenomena, a radically de-sacralizing epoché. But in the later works this epoché has been lifted. If phenomenology remained neutral towards religious phenomena the later "thought of Being" shows a surprising openness to sacral phenomena: the holy, the gods and god. As Kovacs puts it, the essential thinking of the ontological difference of Being, as the lighting-process of Being (as the thinking of Being as further than all beings and yet closer to the human being than any being), remains in a state of theological indifference. Nevertheless, it does not exclude the possibility of a theological difference, i.e., the possibility for a relation of There-being (man) to God (the religious meaning of transcendence).[178] Basing on the fact that Heidegger has articulated the relationship between Being and Dasein in a way which is strikingly like the relationship between God and the soul in mystical literature Löwith traces a "religious motif" in Heidegger.[179]

In his *Introduction to Metaphysics* Heidegger raises the question how the question of Being is related to Theology.[180]

---

[178]G. KOVACS, *The question of God*, 181. To find the closeness between the notion of God and the notion of Being, see, KOVACS, 276, note16. Richardson hints at the question of God when he says: "... There is one problem, however, that we resolutely avoid: the problem of God. This is not because there is nothing to say about God in Heidegger's thinking. On the contrary, it is because there is too much to say for it to be said merely by indirection." *Phenomenology to Thought*, xxviii. See also W.J. RICHARDSON, "Heidegger and God - And Professor Jonas" *Thought*, 40 (1965) 13-40.

[179]K. LÖWITH, *Denker in dürftiger Zeit*, 72.

[180]In Macquarrie's view Heidegger fails to recognize the varieties of Theology and the fact that there are Theologies which really see God's relation to

According to him the question of Being as such is not raised in Theology. It is stuck in the levels of beings and never gets beyond the beings. Heidegger thinks that the God of metaphysics is religiously inadequate. Again in ID he criticizes metaphysics not because it has to do with a being, God, but because it fails to appreciate the distinctive mode of God's Being.[181] The ID implies not that there is no God, but that God needs to be conceived in terms which are appropriate to his mode of Being. In other words Heidegger does not reject God but only the metaphysical conception of God. And Heidegger thinks that a god-less thinking is more open to God than onto-theo-logic. Hence he says: "The god-less thinking which must abandon the god of philosophy, god as *causa sui*, is thus perhaps closer to the divine god. Here this means only: god-less thinking is more open to Him than onto-theo-logic would like to admit."[182]

One can see in this passage how close Heidegger's thought moves to the religious motif.[183] Again as Kockelmans says, 'Heidegger in his inquiry into gods, as the world-structuring realities, and his silence toward the problem of God in no wise confesses polytheism which *eo ipso* would deny Christian monotheism, but is careful not to cross the realm of human experience. Such care is true respect for the supernatural, and thus superhuman realm and is not its denial.'[184]

---

beings much in the same terms as Heidegger visualized it. For instance, he speaks of Tillich for whom God is not a being, not even the supreme being, but Being itself. He claims further that there were also theologians at least from the time of Dionysius the Areopagite, who spoke of God as *hyperousia*, 'Beyond being.' See, J. MACQUARRIE, *Heidegger and Christianity*, 55.

[181]ID. 72.

[182]ID. 72.

[183]For a detailed treatment of the question of God in Heidegger see, G. KOVACS, *The Question of God*; see also V. VYCINAS, *Earth and Gods*.

[184]J.J. KOCKELMANS, *The Truth of Being*, 133.

Heidegger in fact opens up the way beyond the God of metaphysics to the truly divine God, the God of faith and of religious experience. If the categories of the later Heidegger are taken seriously by the theologian, then God is thought in terms of a voice which addresses us, a call which summons us, an event which happens to us. The relationship of the world to God is experienced as a coming to presence (*Wesen*, taken verbally) which either makes His presence felt to us (*An-wesen*) or withdrawn from us (*Ab-wesen*). Being is also the eternal heart of things, because it is the innermost source of their presence, and because it is the perpetual Being that lets all abiding be.[185] It is the omnipresent and the undefiled.[186]

We do not grasp God by our concepts, but rather we are seized by His presence, summoned by His call, overtaken by the event which happens to us. Rather than con-ceiving God, grasping him about, the task of thought is to let God be God, as Meister Eckhart said, from whom Heidegger has borrowed much.[187] In a word: the paradigm of seeing, substance, cause and

---

[185]M. HEIDEGGER, *Erläuterungen zu Hölderlins Dichtung* ,GA 4 (Frankfurt am Main: Vittorio Klostermann 1981), 71.

[186]HD. 71.

[187]J.D. CAPUTO, "Heidegger's God and the Lord of History," *New Scholasticism* 47/4 (1983) 446. Speaking about Eckhart's conception of God, Caputo observes the following: "Eckhart is always looking for ways to get beyond "God," that is to say, what men call God, whether on the basis of metaphysical theology or even of revealed faith. ... Whatever we *know* about God is not God, for that is God *insofar as* he has been under the sway of human knowledge. Whatever we *want* of God is not God, for that is God *insofar as* he has been brought under the human willing. The only way to God - that is, to the truly divine God, what Eckhart sometimes called the Godhead beyond all God - is to shut down the whole operation of knowing and willing, that is to say, to suspend the operations of subjectivity, to disconnect the *ego cogito,* and let God be *God.* And it was of course at that point that Eckhart invoked the word *Ge-lassen-heit,* letting-be. The "truly divine God" is precisely the God Who *recedes* behind everything which is

concept gives ways to hearing, event, call and letting-be. Atheism and theism are replaced by concealment and un-concealment, absence and presence. The God of Metaphysics is given up for the truly divine God. For Heidegger realizes that the way to deal with the transcendent and "simply other" reality (of God or Being) is not to deal with it at all, but to let it deal with us. The only way to gain access to Being itself is to let Being be and to let it address man. The way of the soul to God is to let God take the lead and to let God effect the unity of the soul with God. For it is beyond the powers of the soul to bring about such a unity itself.[188] Hence Heidegger remarks in the postscript: "Being is no product of thinking. On the contrary, indeed, essential thinking is an event (*Ereignis*) of Being."[189]

12. CONCLUSION

Heidegger in explaining his own philosophy as a rethinking of the tradition tries to think Being instead of the 'beingness' of beings by way of overcoming traditional philosophy which in his view is a return to the proximity of the most near. Heidegger's lecture, "*Was ist das-die Philosophie*?" makes it quite clear that the Being-question is the very essence of his conception of philosophy. Philosophy consists in listening to the voice of Being, in the attentive disposition toward the intimations of Being in and through the many beings, in a correspondence with the Being of beings.[190] Philosophy is thinking as the thinking of Being. With the questioning and thinking of Being, we come upon the ontological difference, with its stress of the difference between Being and being, and Dasein's transcendence to the Being of being. The

---

said about him." see, J.D. CAPUTO, *Mystical Elements,* xviii.

[188]J.D. CAPUTO, *Mystical Elements*, 25.

[189]WM. 356.

[190]M. HEIDEGGER, *What is Philosophy*, tr., W. KLUBACK & J.T. WILDE (Albany: NCUP, Inc 1956), 69; 71; 73; 75.

ontological difference is more fundamentally based on the essence of Dasein where the ascent from being to Being occurs. He is named Dasein precisely because he is the *Da* of the *Sein*. His essence is ek-sistence, openness to Being. He responds to the call of Being. He is the shepherd of Being. He guards Being in himself. The Being thus considered in its relation to Dasein we have attempted to analyze from different standpoints.

In its negative significance Being is seen as Non-being while in its positive significance Being is considered as transcendence. Instead of attempting to ground Being in a transcendent being Heidegger tries to ground it in the very transcendence of Being. He finds the principle of ground initially connected with the problem of truth. And truth for Heidegger is the very 'uncovering' of Being. Along with the unconcealment of Being in truth language breaks out where the Being finds its home. Man makes his journey in the Being through the path of language which makes the central concern of our next chapter.

Our discussion also take up the questions of time, world and their inter-relationship to the question of Being. Being and time belong together. Man, though he ek-sists in Being is temporalized. He lives in a world which is no more something merely instrumental but has a dignity of its own. Thinking for Heidegger is the thinking of Being as such. Therefore, he invites us to break out of representational and rationalistic thinking, leaving behind the rules of logic. However the question remains, can one really speak about Being as such without speaking of the Being of beings? This Being is that which gives. It is the groundless ground. The authentic experiencing of this Being which is beyond differentiations includes a movement from the philosophical God in Heidegger's interpretation to the truly divine.

CHAPTER FOUR

LANGUAGE: THE SAYING OF BEING

1. INTRODUCTION

Martin Heidegger has developed his understanding of the nature of language in relation to his philosophical problematic of the question of Being, over two distinct phases of his career. The first of these two phases is to be found in BT and the second is to be found in his post BT works. However, in each phase of his career Heidegger's understanding of language has been founded in his understanding of Being. His understanding of language in the second phase differs from his understanding of language in the first because he has shifted his accent from Dasein to *Sein* (man to Being) as he passes from the first phase of his career to the second.

Heidegger's understanding of language in BT arises out of his investigation of Dasein, the being whose mode of being is to inquire concerning its own Being. In the later works Heidegger has turned from the analysis of Dasein to the more general investigation of Being as such. Accordingly, his understanding of the nature of language, as it appears in the later works, develops out of this new perspective. For most philosophers the crucial problems in philosophy of language lie in the area of semantics- the relation between language and the nonlinguistic features of the world which are somehow made accessible to us through language. Heidegger would not deny that this is a most important area in philosophy of language; but for him there is an even more important problem, the essence or ground of language itself. He notes a kind of impatience in many thinkers with this problem; indeed most do not ask such a question but simply take it for granted that language is essentially some kind of system of symbols for the purpose of communication.

Already in the BT, Heidegger showed dissatisfaction with this kind of treatment of language: "In the last resort, philosophical research must resolve to ask, what kind of Being goes with language in general. Is it a kind of equipment ready-to-hand within-the world, or has it Dasein's kind of Being, or is it neither of these?" [1] Heidegger makes his own way towards the question of language and its implications which is searching and philosophical in orientation. The purpose of this chapter is to investigate Heidegger's understanding of the nature of language in the two distinct phases of his thinking in relation to the elaboration of his philosophical problematic. The two distinct and apparently divergent understandings of the nature of language are reconciled by means of a study of the relationship of the problem of language, as the problematic of the meaning of Being develops over each period as well as over his entire career.[2]

2. ON THE WAY TO LANGUAGE

Heidegger's understanding of language has been intimately associated with his thinking of Being over the entire course of his works. According to Rampley, the origins of Heidegger's concern

---

[1]BT. 209.

[2]For an understanding of the early phase of Heidegger's language we depend chiefly on his major text BT. Since Heidegger's later works encompass a rather large corpus, a definitive treatment of all the texts that touch on the problem of his later understanding of language is not attempted. We limit our analysis to the following principal texts: *An Introduction to Metaphysics, Letter on Humanism, What is called Thinking?, On the way to Language* and *Poetry, Language and Thought*. References are also made to Heidegger's interpretations of Poets like Hölderlin, Trakl, Rilke and others.

with language can be traced back to his work on Duns Scotus, and then back further still, to his doctoral dissertation on *The Theory of Judgement in Psychologism* of 1913, and to his survey of 'Recent Researches into Logic' of 1912. [3] In the early phase of his writing, for example in the *Habilitationsschrift*, the language-question emerged probably as the question concerning logic and the relation of logic to language.[4] In WL Heidegger claims that from the beginning, the question of language and logic was closely related to the Being-question. He states: "... reflection on language, and on Being had determined my path from early on..."[5] Further Heidegger states: "In summer semester of the year 1934 I gave a lecture course under the title "Logic." It was, however, a reflection on *logos*, in which I sought for the essence of Language."[6]

Heidegger does not consider the reflection on the essence of language to be a philosophical endeavour separate from his reflection on the Being-question. Rather he asserts in LH that the thinking of the truth of Being elevates reflection on the essence of

---

[3]M. RAMPLEY "Meaning and Language in early Heidegger: From Duns Scotus to Being and Time," *The Journal of British Society for Phenemenology* 25/3(1994) 209.

[4]For an elaborative study of *Habilitationsschrift* see J.D. CAPUTO "Phenomenology, Mysticism and the "*Grammatica Speculativa*": A Study of Heidegger's "*Habilitationsschrift*," *Journal of the British Society for Phenomenology* 5/2 (1974) 101-117.

[5]WL. 7. Heidegger's writings on language, collected under the title *Unterwegs zur Sprache*, (tr., *On the Way to Language)* are among the last works he has had published. Up until that time there are references to language and its problems scattered throughout all his works. His interpretations of the pre-Socratics deal with the ontological character of the Greek language as a central theme. Heidegger himself contends that the problem of language has always been a major focal point for all his thinking; from the very beginning of his thought he has stressed the kinship between language and Being.

[6]WL. 8.

language to a new rank.[7] In the *Physics* seminar of 1940, Heidegger speaks of the new rank in the reflection on language thus: "... because man is only insofar as he comports himself with beings as such, revealing them and concealing them, man can and must have the 'word,' sc. speak of the Being of beings."[8]

Here Heidegger indicates that in man's special relationship to Being he is active with regard to the revealing and concealing of Being in beings. Man, as engaging in this particular activity, must have a particular capability - the 'word.' This intimation of the role of the 'word' as proposed by Heidegger indicates that speech is directly involved in the disclosure of the Being of beings. In the sentence following the passage quoted above, Heidegger asserts that this role of the 'word' and of speech is a very special one, to be distinguished from the role of everyday language, which is a derivative from the speaking of the original 'word.'

The early Greeks have defined the essence of man as *zoon logon echon* - the living being having language.[9] Speaking, then, is not only natural for man but distinctive of man, in fact so distinctive that man is defined in terms of his speaking. Heidegger interprets the Greek definition of man as follows: The statement means not only that man possesses in addition to other faculties also the capacity to speak. The statement wants to say that first of all language enables man to be precisely that living being which he is as man. As the speaking being, man is: man.[10]

Thus language holds a crucial place in the being of man. Ordinarily, however, we interpret the Greek definition of man not as the "living being that has language" or the " living being that speaks" but as the " rational animal." Heidegger adverts to this

---

[7]LH. 222.

[8]Quoted in translation, W.J. RICHARDSON "Heidegger and the Origin of Language," *International Philosophical Quarterly*, 2 (1962) 404.

[9]BT. 208.

[10]PLT. 189.

later interpretation of the Greek definition of man: "The later way of interpreting this definition of man in the sense of *animal rationale* 'something living which has reason,' is not indeed 'false' but it covers up the phenomenal basis for this definition of 'Dasein.' Man shows himself as the entity which talks."[11]

According to the popular view, language is the audible activity which gives external expression to human thought and feelings.[12] Heidegger analyses the popular view of language in terms of three assertions. First, language is taken to be an external expression, a notion which presupposes the idea of something internal which externalizes itself. The idea of language as something external requires that its intelligibility should be explained retrogressively by appealing to internal ideas and feelings.[13] Secondly, it seems obvious that language is an activity of man. Thus it seems proper to say that man speaks, but not that language does. Finally, human expression is taken to represent real and unreal things. Heidegger does not criticize this popular view of language. But since the popular view ignores completely the oldest natural cast of language, it can never lead to the revelation of language as language.[14] These remarks indicate that it will be Heidegger's strategy to go beyond the traditional view of language by revealing language as a primordial event by which Being is raised to the level of *Logos*. According to Heidegger the inescapable

---

[11]BT. 208.

[12]PLT. 192.

[13]If the concept of language as external expression requires that linguistic intelligibility should be explained by appeal to something internal, then one is lead to wonder whether the intelligibility of internal ideas and feelings does not also require the same kind of retrogressive explanation, and how such regress, once started, could ever be stopped. The only way to avoid unanswerable questions such as these seems to be to abandon the internal-external dichotomy.

[14]PLT. 193.

language with which we are all intimately acquainted is "something different from a mere excitation and spreading of sound."[15]

The external expression is accomplished only by responding to language in a two-fold way: by listening for its call and in replying to it.[16] Again it is the presence of things in the call of language which enables us to re-present them externally as real or unreal. What is important is to learn to live in the speaking of language and to constantly examine our ability to respond to it.[17] Heidegger takes it as his task to talk about language, to discuss it, to dispose it, to bring to its place of being not so much language as ourselves.[18] In order to reach this objective, he uses as his leading clue, the expression "language is language." Although it appears to be an empty tautology, its emptiness is supposed to open up a certain depth.[19] Because language is nothing but itself, Heidegger argues, it is always with us in a fundamental way. Man speaks constantly. Language as 'pure spokenness' is the inescapable thing which enables man to be the living being he is as man.[20]

For analytic philosophy, for linguistics, meta-linguistics, psycholinguistics and the rest language is an "object" of scholarship and calculative thought. But what Heidegger demands is an experience with language.[21] To experience language is to have one's relationship to language transformed. In the experience of language, we discover that we are in the grip of language. We discover in this experience that words let things be (appear). Our

---

[15]PLT. 207.

[16]PLT. 209.

[17]PLT. 210.

[18]PLT. 190.

[19]PLT. 191-192.

[20]PLT. 189.

[21]WL. 57. *mit der Sprache eine Erfahrung zu machen.*

very relationship to language is thereby transformed, for we come to realize that it is not we who speak but language which speaks through us. However it should be noted from the outset that Heidegger's approach to language does not fit into any readily available category or school. As Macann points out 'Heidegger's approach to language was highly idiosyncratic and so has not furnished the basis for a distinctive conception of philosophy or of a distinctively philosophical conception of language. Neither ordinary language nor the science of linguistics was of much concern to Heidegger -which is not to say that Heidegger was not acutely aware of the way ordinary language works in discourse or of the advances that have been made of another kind altogether.'[22] Heidegger himself, when considering the very question of philosophy's definition, has stated:

> We are introduced to and become acquainted with what philosophy is only when we learn how, in what manner, it is. It is in the manner of correspondence which is to the voice of the Being of being. This correspondence is a speaking. It is in the service of *language*. ...it is considered more correct to say that language is in the service of thinking rather than that thinking, as correspondence, is in the service of language.[23]

There is a subtle distinction that is drawn by Heidegger between language and speaking. This is to say that speaking is the existential risk of commitment that transcends language as a system of symbols only ready for use, but not used. And, it is in this sense that Heidegger argues that "philosophical research will have to dispense with the 'philosophy of language' if it is to inquire into 'the things themselves'..."[24] With these discussions we now move on to Heidegger's way of looking at language as has been presented in his master work BT.

---

[22]C. MACANN, ed., *Martin Heidegger: Critical assessments, Vol.III: Language* (New York: Routledge 1992), 6. Hereafter, C. MACANN, *Martin Heidegger*.

[23]WP. 91-92.

[24]BT. 209-210.

3. LANGUAGE IN *BEING AND TIME*

It is section 34 of BT in which Heidegger speaks of language thematically for the first time. It is entitled "Being-There and Discourse. Language." Heidegger says "... language now becomes our theme for the first time ..." and it has "its root in the existential constitution of Dasein's disclosedness."[25] His purpose is to show that the phenomenon of language has its roots in the existential constitution of Dasein's disclosedness. Also as part of the task of displaying the "ontological locus" of language Heidegger gives a preliminary discussion of the role of language in the disclosedness of the 'there' (*Da*)of Dasein.[26] He calls "the existential-ontological foundation of language *Rede*"and defines *Rede* as the 'articulation of intelligibility.'[27] For the existential-ontological foundation of language Heidegger uses the word *Rede* which in everyday German usage means 'talk,' 'speech,' or 'discourse.'[28] It seems that the very choice of the term '*Rede*' points us towards everydayness rather than towards something abstract and 'scientific.' It is an attempt to begin our search for language in the 'everyday' saying and speaking of Dasein rather than when we seek its meaning in a highly sophisticated science of linguistics.

---

[25]BT. 203.

[26]As J. Aler rightly observes Heidegger briefly touches on the theme of language while mentioning the relationship between observation of objects (*Vernehmen des Vorhandenen*) and language. (BT. 85-90) Corresponding to this cursory indication in the investigation of ek-sistence is a passage in the methodological prelude to the investigation: in characterizing the *logos* of phenomenology, Heidegger also considers the relationship between *logos* and *phònè* (BT. 55-58). J. ALER "The Conception of Language in *Being and Time*" in C. MACANN, *Martin Heidegger*, 24-25.

[27]BT. 203-204.

[28]The term *Rede* which Heidegger uses as equivalent to the term *logos*, is being translated variously. Throughout our discussions we make use of the word discourse in place of *Rede*.

Heidegger himself makes the point that language comes to the fore for the first time within the context of communication, and not as some present-at-hand waiting there to be analyzed. So when Heidegger defines *Rede* as 'articulation of intelligibility,' he seems to be saying that language encountered in our everyday communication with others finds itself always already rooted in an articulation of intelligibility. "The intelligibility of Being-in-the-world - an intelligibility which goes with state-of-mind expresses itself as *Rede.*" He adds immediately thereafter: "the totality-of-significations of intelligibility is put into words." And he interprets himself to mean that "to significations, words accrue. But word-things do not get supplied with significations."[29] The central phenomena, therefore, are significations; words come to our lips already pregnant with significations within the context of our sharing what we have pointed out with others.

Language as a totality of words- a totality in which *Rede* has a 'worldly' being of its own; and, as an entity within-the-world, it becomes something which we may come across as ready-to-hand.[30] Language, therefore is a complex phenomenon; it is not merely a totality of words, but a totality in which the *Rede* gets expressed. One does not have language if one merely defines it as a totality of words and leaves it. Language is rather a totality of words in which the intelligibility of our being-in-the world is expressed. Any discussion of language, therefore, while finding language ready-to-hand within the world, must take it as a whole, that is, must advert to the roots of language in the articulation of intelligibility of Being-in-the-world, and not take it simply as a "totality of words." "Language," Heidegger concedes, "can be broken up into word-things which are present-at-hand." Heidegger does not work out an elaboration of language based on

[29]BT. 204.

[30]BT. 204.

a foundation ontologically more primordial. He attempts merely to locate the phenomenon of language in the existential analytic of Dasein and describe its meaning, structure and possibilities with respect to this locus which is essentially time-bound.[31]

In the preparatory analytic of man's Being, one obviously expects also to find an explanation of language. Only relatively late- in the next to the last chapter of the preparatory analytic - is language even mentioned. The analytic concentrates least of all on this anthropological, pre-eminently fundamental phenomenon. It does not deal with language as the range of systematized possibilities of expression by means of symbols which appear in the possible combinations of vocal sounds; it is concerned, rather, with speech as that form of human behaviour in which these possibilities become materialized. In this analytic, linguistic phenomena are dealt with in the same way as other basic forms and principles of man's Being.

As mentioned, Heidegger approaches the problem of the meaning of Being by means of a phenomenological investigation of the existence of Dasein which has the capacity to inquire concerning its own Being. This investigation leads him to maintain that the Being of Dasein is a structural unity constituted according to three fundamental modes of equal importance. Heidegger indicates that the existential-ontological foundation of language is discourse which is existentially equiprimordial with state-of-mind and understanding in Dasein's constitution.[32] Thus under the

---

[31]It is well known that BT is mainly concerned with man's Being in order to lay the foundation for a general ontology. This preparatory reflection is performed in two phases. First, a number of structures of man's Being are developed in an 'analytic,' which in itself is again 'preparatory'; second, these structures are explained as modes of temporality. Temporality is the essence of man's Being; thus, time is comprehended as the horizon of that understanding of Being which is characteristic of man.

[32]BT. 203.

mode of discourse, language is found to be constitutive of the existence of the individual human being. And Heidegger considers discourse to be the ontological ground of language. "The existential-ontological foundation of language is discourse or talk."[33] Accordingly a treatment of Heidegger's language must take discourse as its starting point. Heidegger considers that any other modes of language, i.e., written language, are founded in discourse and derived from it.

In analyzing the Being of Dasein in BT Heidegger has given consideration to the problem of language in three principal areas of the text. First, language as discourse (*Rede*), appears as constitutive of the disclosedness of the 'there' (*Da*) of Dasein. Second, as 'idle talk' (*Gerede*)[34] language enters the problem of the temporal analysis of Dasein's everydayness. Third, language appears as the medium of the function of Dasein's conscience. Language in BT which Heidegger finds to be grounded in discourse is found to be equiprimordial with state-of-mind and understanding - two of the elements which are constitutive of the disclosedness of Dasein.[35] Dasein comes to express itself in assertions. As it makes an assertion about an entity it can pass from its concernful dealing with the entity as ready-to-hand to become an observer of the entity as present-at-hand. Language itself may be conceived as an entity ready-to-hand within the world, with its Being distinct from the Being of Dasein. Accordingly language becomes the object of the science of linguistics.

According to Heidegger the primary mode of discourse is speech or talk. As an essential constitutive element of *Da*, the

---

[33]BT. 203.

[34]The term *Gerede* is again translated differently as prattle, chatter, idle talk etc. We confine to the term idle talk in our expositions.

[35]BT. 172.

openness of Dasein, discourse is basically not the empirical sound-language, but is the articulation and meaningful ordering of Dasein as the to-be-in-the -world. By being in the world we have a mood-like understanding of our to-be-in-the-world. Our to-be-in-the-world, the openness of Being, is revealed to us in our 'thrownness.' It is not chaotic but comes to us in meaningful structures. These structures-already implied in our 'thrownness'-we bring up through our projection in our words wherein we preserve them. Discourse, as the articulation of these meaningful structures of that which is, primarily is the language of Being, the self-utterance of Being, and not of human speech. By listening to discourse, we speak. Language is thus in every respect constitutive of our ek-sistence. That it is an ontological derivative phenomenon by no means excludes this fact. But the mode of ek-sistence in which Heidegger's exposition reveals the constitutive character of language, is, within the general perspective of the preparatory analytic, the average everydayness: language is a tool to be used in social interaction.

4. *REDE*: THE LANGUAGE OF DASEIN

In accordance with the phenomenological slogan, to which Heidegger himself initially adhered- 'To things themselves', language plays a relatively subordinate role in the first phase of Heidegger's philosophy, i.e., as the articulation of intelligibility. The term *Rede* (discourse) which Heidegger adopts in to talk about language belongs within the priority Heidegger accords to *praxis* over theory (*parole* before *langue*).[36] As we have already

---

[36]C. Macann opines that even the more theoratically appropriate term assertion (*Aussage*) is still connected with the communication of meaning through language and is directed toward a pointing out (*aufzeigen*) which is itself linked to the term employed to characterize the manifestation of being: to show (*sich zeigen*) itself. The pre-lingual character of language as *Rede* (discourse) is moreover linked to the extra-lingual character of *Rede* as *logos* in this respect; Heidegger is still thoroughly Husserlian, assuming an

seen in BT Heidegger acknowledges man as primordially Being-in-the-world, within the context of which language as *Rede* emerges as the 'articulation of the intelligibility of Being-in-the-world.' Language emerges as being co-constitutive of the disclosedness of Dasein, of Being-in-the-world, a Being-in-the-world working out with others its possibilities from out of the state of throwness in which it finds itself. And hence according to BT, the locus of the phenomenon of language is in Dasein as Being-in-the-world.

In his analysis of the structure of the disclosedness of the 'there' of Dasein Heidegger includes discussions of the three constitutive items of Dasein's disclosedness, namely, state-of-mind, understanding and discourse. Heidegger's treatment of language in this early stage of BT appears to indicate that language is more closely allied to understanding than to state-of-mind.[37] Understanding founds interpretation, which in turn founds assertion. These latter two find discourse as their means of articulation. State-of-mind comes to bear upon discourse inasmuch as state-of-mind maintains itself in a certain understanding to which there corresponds an interpreting.

While analyzing understanding Heidegger indicates that the world is 'there' as that for the sake of which Dasein is. He continues: "In the 'for-the-sake-of-which,' existing Being-in-the-world is disclosed as such and this disclosedness we have called understanding."[38] Understanding thus defined, grounds "significance" as "that on the basis of which the world is disclosed

---

essential meaning core with reference to which language becomes an incidental or even an accidental overlay, a tool employed for a given purpose, that of speaking forth (*Heraussage*) and communicating (*Mitteilung*) - language as the overt expressedness of *logos.* C. MACANN, "Introduction" in *Martin Heidegger*, 7.

[37]BT. 203.

[38]BT. 182.

as such."[39] Thus it is by way of the understanding that Dasein's world and its totality of involvements is disclosed to it. Dasein as understanding projects its Being upon possibilities. But, the projecting of the understanding has as its own possibility of working its own possibility out. Heidegger gives this development of the understanding the name "interpretation"(*Auslegung*). Accordingly, "interpretation" is grounded in the understanding.

To Heidegger, "Meaning is that where the intelligibility of something maintains itself" and is "that which can be articulated in a disclosure by means of which (Dasein) understands."[40] In other words meaning indicates the fact of Dasein's understanding of the Being of entities within-the-world, which entities Dasein has come to understand. Accordingly meaning is characteristic of the Being of Dasein and is not attached to entities somehow lying behind them. Only Dasein can be meaningful or meaningless. Accordingly Dasein articulates the meaning of something. In this sense articulation is a concrete selling out of what Dasein, as understanding, has interpretatively sketched out beforehand with regard to the Being of an entity.[41]

Dasein, man as ontologically considered, "carries in its own most Being the character of not being closed off."[42] Being-in-the-world is a fundamental unitary phenomenon and Heidegger seeks to elaborate the Being-in dimension of this unitary phenomenon. Hence he sets out to discuss the Being-in of Dasein; he wishes to discuss man as disclosedness, as the being who opens up a world of its concern. The constitutive ways of being the "there" of being open and of opening up a world are according to Heidegger state of mind (*Befindlichkeit*), understanding (*Verstehen*) and discourse (*Rede*). They are all constitutive moments equiprimordially of the

---

[39]BT. 182.

[40]BT. 193.

[41]BT. 195.

[42]BT. 171.

unitary phenomenon of Being-in-the-world, which is a Being-with-others. In this way we shall see the ontology of language within the fundamental ontology of Dasein. By 'intelligibility-articulation' (*Rede*), Heidegger means: the making manifest or bringing out into the open of what is to be discussed in such a way that the envisaged articulation emerges from the very content of the articulatable itself. Intelligibility -articulation is not language, and is thus neither written nor spoken, but it is that which makes language possible.

In BT, Heidegger maintains that Dasein's Being-in-the-world involves Being-with others, wherein Dasein's solicitude for the other finds its everyday expression in communication. This communication takes the form of conversation, which involves four constitutive elements: what the conversation is about, what is said in the conversation, the listening within the conversation, and the sharing of the conversation.[43] As Walter Biemel observes, in BT this distinction is only briefly mentioned; later, particularly the moment of manifestation (in the sense of uncovering or freeing) comes more and more to the fore, and this element then leads to the connection between *logos* and truth.[44]

Heidegger asserts in BT, that everyday speech gives predominance to the temporal ec-stasis of the present.[45] It is

---

[43]BT. 206.

[44]W. BIEMEL "Poetry and Language in Heidegger" in J. KOCKELMANS, ed., & tr., *On Heidegger and Language* (Evanston: Northwestern University Press 1972),72.

[45]Since discourse expresses itself factically in language, the present as making-present has a privileged constitutive role with regard to everyday speech. Heidegger also maintains that the fundamental problems of the science of linguistics can be validly explored only if the problem of the relation of Being and truth is treated in the light of the problematic of temporality. Heidegger promises to handle these problems in BT, but he published this work incomplete, and the problems of the constitution of language remain untreated in BT.

within language that Dasein understands its Being as 'care'- as thrown into its current situation and projecting its own concrete possibilities. What the discourse is about is either an entity within the world or a state-of-mind of Dasein. What-is-said-in-the-talk is the verbal expression which serves as the means of communication from one person to another. The making-known is the expressive sharing of an understanding between Dasein and the other. Heidegger thinks that the traditional conceptions of the essence of language have been founded upon one or another of these four items or upon a combination of them. But the essence of language is not to be found in any of the items or even in the sum total of them. Instead what is decisive for the problem of the essence of language is "to work out in advance the ontologico-existential whole of the structure of discourse on the basis of the analytic of Dasein."[46]

That which the discourse is about is the meaning or intelligibility which concernful Dasein encounters within the world. When entities within-the-world have come to be understood, they have 'meaning' (*Sinn*). Thus it is 'meaning' which is articulated in discourse. For Heidegger "Meaning is that wherein the intelligibility of something maintains itself."[47] Hence discourse is the means whereby meaning is explicitated. In short, discourse is that wherein intelligibility is found to be as that which it is. Discourse as the articulation of the intelligibility of the disclosedness of the 'there' is a fundamental way of Being of the 'there.' Inasmuch as disclosedness as constituted by the Being of Dasein as Being-in-the-world, is worldly, so that discourse will be worldly.[48] Accordingly discourse is to be distinguished from language, inasmuch as discourse is the ontological foundation of

---

[46]BT. 206.

[47]BT. 192-193.

[48]BT. 204.

language. However discourse gets expressed in language which is a totality of words and which is used by Dasein in its solicitous Being-with others.[49] Heidegger states:

> In any talk or discourse ... something is said as such whenever one wishes, asks, or expresses oneself about something. In this 'something said,' discourse communicates.[50]

Dasein's everyday understanding is remote from authentic understanding. This is so because Dasein understands what is said in the talk as such, i.e., what the words of a conversation signify, and understands the purpose of the talk only approximately and superficially. For an average intelligibility permeates spoken language with the result that Dasein loses sight of a primordial understanding of Being and is primarily concerned with what is said in the talk. In what-is-said-in-the-talk, Heidegger indicates that the intelligibility has always been articulated or spelled out before concernful Dasein gives an appropriative interpretation of it.[51] Thus intelligibility as that which is articulatible in discourse is identified by the term 'totality-of-significations'. Inasmuch as intelligibility expresses itself as discourse, the totality-of-significations, understood as the articulation of intelligibility, is put into words. Accordingly significations are the primary bearers of meaning and words are supplied to significations in order that significations may be expressed. Heidegger states: "To significations words accrue. But word-Things do not get supplied with significations."[52] In this way discourse gets expressed in language.

Heidegger asserts that Dasein as Being-in-the-world maintains itself in some definite way as concernful being-with-one-another. Dasein expresses itself according to the fulness of its

---

[49]BT. 204.

[50]BT. 205.

[51]BT. 203.

[52]BT. 204.

disclosedness as Being-in-the-world. Heidegger states: "Such Being-with-one-another is discursive as assenting or refusing, as demanding or warning, as pronouncing, or interceding, as 'making assertions,' and as talking in the way of 'giving' a talk."[53] When the other speaks Dasein also hears the message of the other. Only where talking and hearing are existentially possible can Dasein harken. Because talking and hearing are based upon understanding, it is only on the basis of Dasein's potentiality-for-Being as hearing that anything like harkening becomes possible. Heidegger says about listening: "Listening to ... is Dasein's existential way of Being-open as Being-with for others. Indeed, hearing constitutes the primary and authentic way in which Dasein is open for its own most potentiality-for-Being ..."[54]

The fourth constitutive item of discourse the 'making known,' occurs within communication in such a way that "the articulation of Being with one another understandingly is constituted."[55] Through communication a "co-state-of-mind" and an understanding of Being-with get shared, because Dasein as Being with others is already manifest in a "co-state-of-mind" and in a "co-understanding." But in discourse Dasein appropriates (*zueignen*) Being with, and Being-with becomes explicitly and expressively shared.

The four items constitutive of discourse "are not properties which can just be raked up empirically from language."[56] They "are existential characteristics rooted in the state of Dasein's Being, and it is they that first make anything like language ontologically possible."[57] This is what Heidegger means when he says that anything like a science of linguistics is founded on a

---

[53]BT. 204.

[54]BT. 209.

[55]BT. 205.

[56]BT. 206.

[57]BT. 206.

prior ontology of language rooted in the existential analytic of Dasein.

The analysis of language and discourse in section 34 of BT seems to be tentative in nature.[58] This is evident from the latter part of section 34 which Heidegger concludes with a series of questions relating to the nature of language, where he says:

> ...what kind of Being goes with language in general. Is it a kind of equipment ready-to-hand within-the-world, or has it Dasein's type of Being, or is it neither of these? What kind of Being does language have, if there can be such a thing as a 'dead' language? What do the 'rise' and 'decline' of a language mean ontologically? ... Is it an accident that proximally and for the most part significations are 'worldly,' sketched out before-hand by the significance of the world, that they are indeed often predominantly 'spatial'? Or does this 'fact' have existential-ontological necessity? and if it is necessary, why should it be so?[59]

A language always presupposes a certain interpretability of Dasein's understanding which is discourse. In his later works Heidegger goes deeper into the problem of language and clearly states that discourse is the language of Being and that existentially interpreted meaning primarily belongs to Being. Our understanding as well as our language is a response to the discourse of Being.

5. *GEREDE*: EVERYDAY LANGUAGE

Dasein is at home in a world and finds the meaning of the things he makes and uses within the context of this world. Dasein lives in communication with others who share the same world

---

[58] *Rede* is existentially language. At the same time *Rede* is existentially equiprimordial with state-of-mind and understanding. Heidegger himself, it seems to me, is not very explicit about the specific function of *Rede* as co-constitutive of the disclosedness of Dasein. With Heidegger's Japanese friend Tezuka we must admit that Heidegger's discussion of language in this chapter of BT remains so sparse. (WL.41).

[59] BT. 209.

with him. The language of Dasein embodies, as it were, and makes explicit the intelligibility of Dasein's Being-in-the-world. It is a language more primordial than the theoretical and abstract language which articulates the intelligibility of things in their present-at-hand kind of Being. So what is searched for is an everyday kind of language and not the sophisticated language of scientific man; an every day concrete speaking language and not a calculating language. It is in this context that Heidegger speak of *Gerede* (idle talk) in section 35 of BT. In such everyday discourse the hearer does not allow himself to be brought "into a Being-towards what the discourse is about so as to have a primordial understanding of it."[60]

> And because this discoursing has lost its primary relationship of Being-towards the entity talked about, or else has never achieved such a relationship, it does not communicate in such a way as to let this entity be appropriated in a primordial manner, but communicates rather by following the route of gossiping and passing the word along.[61]

According to Heidegger the lack of grounds prevalent in Dasein's average understanding becomes aggravated to a complete groundlessness. The everyday Dasein's conversation becomes merely gossip and passing the word along. Hence Dasein's conversation develops to what Heidegger calls idle talk. Idle talk is not a disparaging term but "signifies a positive phenomenon which constitutes the kind of Being of everyday Dasein's understanding and interpreting."[62] Idle talk closes Dasein off from its being a Being-in-the-world and covers up entities within the world.[63]

Everyday Dasein as it engages in idle talk is cut off from the

---

[60]BT. 212.
[61]BT. 212.
[62]BT. 211.
[63]BT. 213.

authentic understanding of its disclosedness. In other words the possibility of projecting itself authentically is overlooked by Dasein as it remains engrossed in idle talk. By maintaining itself in idle talk Dasein cuts itself off from other aspects of its authentic Being-in-the-world. Heidegger states: "Ontologically ... when Dasein maintains itself in idle talk, it is as Being-in-the-world - cut off from its primary and primordially genuine relationships-of-Being towards the world, towards Dasein-with and towards its very Being-in."[64]

Accordingly the phenomenon of idle talk reveals that everyday Dasein can and does cut itself off from its own authentic way of Being as Being-in-the-world. Speaking on the question of idle talk V. Vycinas observes the following: 'In the everyday usage of language, the speaker and the listener are not oriented to the discourse, but they merely stick to the spoken words and their combinations separated from their ontological basis (discourse). Thus, the *dictum* replaces reality and becomes the location of truth. Such an eradicated discourse, Heidegger calls chatter (*Gerede*). Just as discourse means the disclosure (openness and also truth) of Dasein, so chatting means the opposite, namely, the concealment (untruth) of Dasein.'[65]

To put in other words, in discourse, Dasein does not share an entity with the other, he shares an understanding of the entity leaving the entity out of the grasp of the other. Thus, in order to insure a proper sharing of what the discourse is about, Dasein must engage in authentic hearing and speaking. If authentic speaking and hearing do not characterize a discourse, inauthentic communication and misunderstanding result. Dasein's inauthentic communication develops into gossip, chatter, and idle talk with the result that Dasein is buried in the domination of the average

---

[64]BT. 214.

[65]V. VINCINAS, *Earth and Gods,* 49.

everyday ways of the 'they.' Through the discourse of conscience Dasein extracts itself from the domination of 'they' and comes into the authenticity of its Being as resolute. Dasein understands the call of conscience by choosing to want to have a conscience and responds to the call of conscience in the mode of keeping silent which Heidegger calls reticence. The participants in the conversation can properly understand what the conversation is about only by listening in silence to what is being said in the conversation. Accordingly the Heidegger of BT identifies silence as a proper mode of discourse in the inauthentic mode of everyday conversation. However, silence is not restricted to be a mode of everyday conversation.

Now, idle talk is the possibility of understanding everything without previously making the thing one's own. If this were done, idle talk would founder; and it already guards itself against such a danger. Idle talk is something which anyone can rake up, it not only releases one from the task of genuine understanding, but develops an undifferentiated kind of intelligibility, for which nothing is closed off any longer. T. Langan distinguishes talk and idle talk as expressions of authentic and in-authentic modalities of Being-in-the-world. He says:

> ...discourse allows the synthetic transcendence of subjectivity into objectivity by the person's encounter (as subject) with an external objectivity that suggests the person's objectivity. Thus, existence as both subject and object is the primordial or authentic status revealed by talk or discourse. However, idle talk is more of a probability in our encounters within the world. ... Idle talk reveals an experience in the absence of perception; it is an appearance- intelligible, yet lacking complete meaning. Idle talk is thus an in-authentic form of communication that does not convey authentic existence.[66]

---

[66]T. LANGAN, *Speaking and Semiology: Maurice Merleau-Ponty's Phenomenological Theory of Existential Communication* (New York: Mouton de Gruyter 1991), 39. However it is to be noted that Heidegger never makes distinctions such as the subjective and the objective.

6. DISCOURSE AND CONSCIENCE

Heidegger has developed the phenomenon of conscience in BT as a mode of discourse - the call of conscience (*Gewissensruf*).[67] He considers the call of conscience as a primordial kind of discourse for Dasein.[68] Through the treatment of conscience Heidegger points to the fact that discourse, as intrinsic to the constitution of the disclosedness of Dasein is ontologically prior to Dasein's everyday communication with others. Heidegger claims that Dasein exists already as a potentiality-for-Being-its-Self, and that this potentiality is attested to by that which is familiar to Dasein in its everyday interpretation of itself as the "voice of conscience."[69] Accordingly, the phenomenon of conscience exists only in Dasein's kind of Being and manifests itself only through Dasein's factical existence.

Inasmuch as conscience brings Dasein before its authentic potentiality-for-Being, conscience gives Dasein something to understand. Thus, conscience discloses and accordingly belongs to the disclosedness of Dasein. Since conscience is revealed as a call, it is a mode of discourse.[70]Dasein responds to the call of conscience by keeping silent [71] and in understanding the call of conscience Dasein understands its own authentic potentiality-for-Being, because "the call comes from the soundlessness of uncanniness, and the Dasein which it summons is called back into the stillness of itself, and called back as something that is to become still."[72]

Accordingly Heidegger concludes: "Thus the mode of Articulative discourse which belongs to wanting to have a

---

[67]BT. 314.

[68]BT. 342.

[69]BT. 313.

[70]BT. 314.

[71]BT. 342.

[72]BT. 343.

conscience is one of reticence."[73] Dasein engages in discourse by hearing as well as talking. As Being-with we could not be a "conversation" if we could not "hear" one another. Hearing is constitutive for discourse. Heidegger adds: "listening to is Dasein's existential way of Being-open as Being-with for others."[74] While hearing is the first existential possibility of discourse keeping silent is another essential possibility of discourse in Heidegger's view and he finds the same existential foundation for it.[75]

Dasein, as Being-with which understands, listens to others. As Dasein loses itself in the idle talk of the 'they'(*das Man)*, it fails to hear its own self in listening to the 'they.'[76] But Dasein's hearing of the voice of conscience interrupts the inauthentic listening to the 'they,' because the call which Dasein heeds appeals to it without mediation. Since the call is unmediated, Dasein itself gives the call.[77] Thus, Dasein as the one who makes the call of conscience, passes over the 'they' to its own self as Being-in-the-world, and relegates the 'they' to a position of insignificance in the discourse of conscience.

7. LANGUAGE: THE SHIFT OF EMPHASIS

We have seen that in BT Heidegger acknowledges man as primordially Being-in-the-world, within the context of which language as *Rede* emerges as the articulation of the intelligibility of Being-in-the-world. Language emerges as being co-constitutive of the disclosedness of Dasein, of Being-in-the-world, a Being-in-the-world working out with others its possibilities from out of the state of throwness in which it finds itself. And hence in BT the

---

[73]BT. 342.

[74]BT. 206.

[75]BT. 206.

[76]BT. 315.

[77]BT. 316.

locus of the phenomenon of language is in Dasein as Being-in-the-world. However as Heidegger notes even in BT every ontology, even the ontology of Dasein, "remains naive and opaque if in its researches into the Being of entities it fails to discuss the meaning of Being in general."[78] It implies that the discussions in BT about the locus of language succeeded only in explicitly locating the phenomenon of language.

Heidegger poised to locate the phenomenon of language for us in Dasein's state of Being. But Heidegger's later works show a considerable change in his understanding of the question of language. It points to a 'turn' also in the realm of language.[79] In BT Heidegger is on the way "toward the concept of Being by way of an interpretation of a certain special entity, Dasein ..."[80] The point that needs emphasis here is that Heidegger is on the way in BT, i.e., BT is a preparation for the revelation of Being. In BT Heidegger makes an analysis of man precisely as that privileged being among beings in that its very Being is an issue for it.[81] Man ontologically considered is called Dasein because of its proximity to Being. Its special character is to be open to Being; it is especially defined with respect to Being. Thus it follows that Dasein is only fully disclosed in the light of Being. Heidegger himself says so explicitly: "The question of *Being and Time* is decisively fulfilled in the thinking of the Reversal. He alone can fulfill who has a vision of fullness. This fulfilment likewise furnishes for the first time an adequate characterization of There-

---

[78]BT. 31.

[79]The reversal (turn) is generally understood to be the shift of emphasis from the earlier writings of Heidegger in which the focus of his attention was on the existential analysis of Dasein to the later writings in which the focus of his attention is on the revelation of the meaning of Being in general.

[80]BT. 63.

[81]BT. 32.

Being, sc. the essence of man ..."[82] This would mean that the very discussion of the locus of the phenomenon of language in the state of Being of Dasein as described in BT is itself provisional and awaits "fulfilment" in the reversal that focuses on the revelation of Being.

The analysis of BT as a whole (the existential analytic of Dasein - fundamental ontology) as J. Aler points out is itself 'provisional, i.e., awaits its fulfilment in the revelation of the meaning of Being in general. This paves the way for a shifting of the explicit focus of our concern from Dasein with emphasis on the "*Da*" to "*Sein*"- a seeking for the revelation of the meaning of Being in the light of which our very analysis of Dasein will be fulfilled which in turn will have an impact on the meaning of language. Hence what Heidegger says about the locus of the phenomenon of language is pro-visional and awaits a fulfilment in the revelation of the meaning of Being.[83] Therefore, ultimately the locus of the phenomenon of language must be found within this fundamental quest of Heidegger.

It is a fact that what is central for Heidegger's thinking remains throughout, i.e., the circle of Being-being quest, so also the continuity of Heidegger's quest is maintained. Yet the focus of Heidegger's attention shifts from an analytic of Dasein which was to prepare us for the revelation of the meaning of Being itself to that revelation itself. As Richardson puts it:

> The characteristic of this change can be stated easily enough. The purpose of SZ was primarily to pose the Being-question, but in the event it proved to be principally an analysis of the Being-Comprehension of There-being, and Being itself was seen in this perspective; Being (the world) was considered basically as the project of there-being. In WW, however, the focal point of Heidegger's

---

[82]Heidegger's "Preface" to W.J. RICHARDSON, *Phenomenology to Thought*, viii-xx.

[83]J. ALER, "The conception of language in *Being and Time*," in C. MACANN, *Martin Heidegger*, 24.

reflection passes subtly from There-being to Being itself.[84]

Heidegger it seems is guided by the overriding concern for the revelation of the meaning of Being itself, though never divorced from its "*Da*." This brings to fore the consequent shift in the primary locus of the phenomenon of language from the "*Da*" of *Sein* to *Sein* and eventually to *Ereignis*. At the end of his discussion of language in section 34 Heidegger observes that the whole purpose of this discussion was "merely to point out the ontological 'locus' of this phenomenon in Dasein's state of Being. By implication then he wishes to tell us that he has not tried to work out the question of language thematically. Richardson notes that 'the third component of disclosedness (*Rede*) is less emphasized in BT than it will become in the later Heidegger.' And he hastens to add that "Heidegger himself is very obscure in this, the least satisfying section, perhaps in all *Sein und Zeit*."[85]

Concomitant with the turn, the later Heidegger's orientation toward the Being-question and attendant problems such as the role of man in relation to Being, the essence of language, and the signification of the Greek concept of *logos* has changed. With regard to the problem of language, a change in Heidegger's position is also evident. In BT language as discourse is constitutive of the disclosedness of Dasein. In the discourse of conscience Dasein is found to discourse solely with itself alone in the mode of reticence. Dasein has language inasmuch as Dasein submits to the necessity of language as constitutive of the disclosedness of the "there."

However the analysis of BT never directly confronts the problem of the nature of language as such. The later Heidegger moves to resolve this difficulty, as he asserts in LH that the essence of language is "that it is the house of the truth of

---

[84]W.J. RICHARDSON, *Phenomenology to Thought*, 238.

[85]W.J. RICHARDSON, *Phenomenology to Thought*, 66.

Being."[86] As man dwells in the clearing of Being he attends thoughtfully to the call of Being. The call of Being which precedes all verbal utterance calls upon man to respond to Being in the equally wordless mode of listening. By responding thoughtfully in silence to the call of Being man becomes manifest as that being who thinks. The call of Being indicates that man's thoughtful nature resides within the essence of language, which is the primal Saying of Being. The role of language is taken up within Being itself. "Language is the clearing-and-concealing advent of Being itself."[87]

Heidegger's understanding of the Greek concept of *logos* has also changed. In BT Heidegger recognized that the ancient Greek thinkers, Parmenides and Heraclitus, held a different understanding of the *logos* than did Plato and Aristotle. But his usage of the term *logos* accepts the derivative grammatical and assertive significations given to the term by Plato and especially Aristotle. In his later works Heidegger takes *logos* to signify a primal collecting or gathering. Man's thoughtful dwelling within the essence of language takes the fundamental form of a 'gathering' of that which lies before him according to the manner in which it lies before him. Heidegger has derived his understanding of 'gathering' as the fundamental function of language on the basis of his interpretation of the pre-Socratic usage of the word *logos*.

Logos which also signified 'language' to the Greeks, indicated primarily the manifestness of beings in Being such that they lie open to the gathering function of essential language as it provides that naming word. By their names things are known as that which they are. This signification he attributes to the thinking of Heraclitus and Parmenides as their thinking can be discerned in

---

[86]LH. 223

[87]LH. 230.

their extant fragments.[88] In ID Heidegger associates the *logos* with the essence of language, as he asserts that the *logos* "contains within itself the origin of the character of all language."[89] Language is not 'the means to portray what already lies before one,' but rather it 'grants presence - that is Being - wherein something appears as existent'[90]

Since language is 'the house of Being,' one reaches Being by constantly going through this house. 'Whenever we go to the well, or walk through the wood, we are always already going through the word "fountain" and the word "wood," even though we are not saying these words or thinking of anything linguistic.'[91] In saying this Heidegger means that language is correlative with experience of the world. However, the central perspective in Heidegger's thought is that of viewing language in relation to Being- that is, questioning why language can be 'the presencing of Being- through returning to the truth of Being which constitutes the origin of language. For as Richardson points out:

> Language takes its origin, then, along with the irruption of There-being, for in this irruption language is simply Being itself formed into word. It follows that in the rise of language, as in the emergence of the There itself, Being retains its primacy. Language as the Over-powering itself, is a might within which the There finds itself, that dominates the There through and through and must be subdued by it.[92]

As indicated in BT the phenomenon of language has its roots in the existential constitution of Dasein's disclosedness ... and if disclosedness is primarily constituted by Being-in-the-world, then discourse too must have essentially a kind of Being which is specifically worldly.[93] In this passage language under the mode of

---

[88]IM. 105-106.

[89]ID. 69.

[90]WL. 146.

[91]PLT.132.

[92]W.J. RICHARDSON, *Phenomenology to Thought*, 293.

[93]BT. 203-204.

discourse is perceived as intrinsic to the structure, or constitution of Dasein and operates within the scope of Dasein's way of Being-in-the-world as disclosedness (*Erschlossenheit*). In the later works where the question of the Being of man is no longer pre-eminent with regard to the Being-question, language is not rooted exclusively in the Being of man. Rather language finds its basis in the clearing of Being. In LH Heidegger speaks of language as the clearing-and-concealing advent of Being itself:

> In its essence, language is not the utterance of an organism; nor is it the expression of a living thing. Nor can it ever be thought in an essentially correct way in terms of its symbolic character, perhaps not even in terms of the character of signification. Language is the clearing-concealing advent of Being itself.[94]

Man's relationship to language is also altered, so that language now embraces man in the truth of Being. In LH, again: 'the point is to think of the essence of language in its correspondence to Being and what is more, as this very correspondence, i.e., the dwelling of man's essence. Man, however, is not only a living being, who besides other faculties possesses Being, wherein living, man ex-sists, while he, guarding it, belongs to the truth of Being.'[95]

Further in LH, Heidegger discusses what he calls the nearness of Being (*die Nähe des Seins*) and proceeds to identify this nearness with the *Da* of Dasein. This nearness, he then insists, realizes itself (*west*) as language.[96] What is obviously suggested is that language is not, as in BT, a mere derivative of the third of the three constituents of the *Da*. Language alone, it seems, is now regarded as enabling Dasein to be its *Da*. Much the same is suggested in WL when Heidegger writes that 'Language first

---

[94]LH. 230.

[95]LH. 237.

[96]LH. 223.

enables man to be that creature which he is as man.'[97] As J. Sallis rightly observes, according to BT, language is the way in which discourse, the articulation of intelligibility, gets expressed. There is virtually nothing to suggest that such expression is anything other than an activity of man, something accomplished by man. Language it seems, is simply an activity of man. But in the later writings, to state it in the boldest fashion, 'language is the language of Being,' and man is called only to respond to 'the unspoken word of Being.'[98]

This would mean that later Heidegger considers the nature of language to be distinct from any specific instance of human speech. Nevertheless the essence of language is related to human speech, because language needs man in order to bring the message of the call of Being into expression. He indicates that all human speech resides within the realm of the essence of language. However most human speech falls short of the call of Being, because it fails to bring that which remains unsaid into expression. Language is always the spoken language of a people. The individual spoken language predisposes its people according to its own modes of speech to respond to it by corresponding modes of thought. At the same time the nature of language lies concealed wordlessly behind the utterance of human speech and reveals itself only in the breach of the failure of the spoken word. Now with these changed perspectives in mind we approach Heidegger's understanding of the nature of language in his later works.

8. THE LANGUAGE SPEAKS

The manner in which Heidegger treats the question of the nature of language indicates that his understanding of the nature of language is unequivocally embedded in the problematic of the

---

[97]WL. 112.

[98]J. SALLIS, "Language and Reversal" in C. MACANN, *Martin Heidegger*, 193.

Being-question itself. Language is very intimately bound up with Being. In IM the fundamental role of language is mentioned as follows:

> ... let us ... attempt to regain the unimpaired strength of language and words; for words and language are not wrappings in which things are packed for the commerce of those who write and speak. It is in words and language that things first come into being and are.[99]

By this strong assertion, Heidegger indicates that language does not occupy a mere secondary or subsidiary role with relation to the Being of things. On the contrary, language is the agency of the coming into presence of that which is. We find this same line of thought in the essay *Words*: "The word's rule springs to light as that which makes the thing be a thing. The word begins to shine as the gathering which first brings what presences to its presence."[100]

Accordingly the function of language is a primal rule which gathers that 'which is' into presence as a thing. Thus, the relationship between thing and word is not a mere scientific or linguistic correspondence. For Heidegger this correspondence between word and thing is founded in man's relationship with Being. In WP he states: "What appeals to us as the voice of Being evokes our correspondence."[101] The principal mode of correspondence is thus a response by man evoked by the appeal of the voice of Being, as this appeal is directed toward man. The relationship of correspondence is evidenced in language and determines the nature of the relationship which language has with the Being of beings. "Being as such determines speaking in such a way that language is attuned ... to the Being of being."[102]

In fact the correspondence, which is attuned to the voice of

---

[99]IM. 113.

[100]WL.155.

[101]WP. 77.

[102]WP. 77.

the Being of beings, is a speaking which is in the service of language.[103] Nevertheless, man is not always authentically open to the correspondence with Being, as Heidegger states: "But only at times does it (correspondence) become an unfolding attitude specifically adopted by us."[104] In LH Heidegger equates the correspondence of language with the dwelling of man's essence: "... the point is to think of the essence of language in its correspondence to Being and what is more, as this very correspondence, i.e., the dwelling of man's essence."[105] Thus the essence of language is man's correspondence to Being. Language, understood as man's dwelling, is regarded by him as the house of Being. Man lives in this house of Being as its guardian. In the essay *What are Poets For?*, language is the means whereby man grasps the Being of beings. It is because language is the house of Being, that we reach what is by constantly going through this house. ... all beings, each in its own way, are our beings in the precinct of language.[106]

Heidegger says that the coming to pass of the ontological difference takes place in language's saying; it is only in the saying of language that Being addresses itself to man. Thus, it is in language's saying that Being 'thinks' and 'speaks.' Being is now no longer merely the coming to pass of Truth as unconcealment; it is now equally *logos* in the full sense of the term. Thus for Heidegger language assumes a much closer proximity to Being itself; it becomes the "house of Being" rather than a mere constitutive structure through which Dasein's Being-in-the-world is articulated. It would thus seem that language is rethought in such a manner that it becomes the concrete presence of Being

---

[103]WP. 93.

[104]WP. 75.

[105]LH. 237.

[106]PLT. 132.

rather than an articulation[107] of transcendental structures.

T. Kotoh observes that after the so-called turning (*Kehre*), corresponding to Dasein's being seen from the perspective of the destiny of Being as the *Da des Seins,* the true being of language is vividly characterized as an echo or response to 'the soundless voice of Being.'[108] Here language speaks as the peal of stillness in that the command of the dif-ference calls world and things into the simple one fold of their intimacy.[109] Even though the openness of Being eventually manifests as language, it is itself the source of language and as such is not under our control. Nevertheless it belongs to the closest neighbourhood of man's being.[110] Hence, Heidegger brings language and Being into the closest proximity. In his view "we are, then, within language and with language before all else. A way to language is not needed. Besides, the way to language is impossible if we indeed are already at that point to which the way is to take us."[111]

Heidegger begins his essay *The Language* with the words: Man speaks (*Der Mensch spricht*). Speaking then is the proper context within which to encounter language. The way to language then, the way towards an experience of its essence, the way towards thinking language in its own terms opens up before us when we put ourselves into the position of 'listening,' hearkening to what is proper (what is the ownness) of language. For

---

[107]The term *articulation* unavoidably suggests that language is a "vehicle" or "instrument" or even an "external sign" for some kind of recessed meaning. This may falsify what would appear to be a much closer connection between thinking and speaking, language and meaning, in the philosophy of the latter-day Heidegger.

[108]T. KOTOH, "Language and Silence: Self-inquiry in Heidegger and Zen" in C. MACANN, *Martin Heidegger,* 47.

[109]PLT. 207.

[110]PLT. 189.

[111]WL. 112.

Heidegger "Language speaks. Man speaks in that he responds to language. This responding is a hearing. It hears because it listens to the command of stillness."[112]

9. EXPERIENCING THE LANGUAGE

In the essay *The Way to Language* Heidegger states: "We are ... within language and with language before all else."[113] Further in the same essay he considers language as something that provides a determinative characteristic of human nature. "The ability to speak is what marks man as man."[114] In as much as language grants to man this capacity of speech Heidegger claims that language is "the foundation of human being."[115] Language is the foundation of human Being according to the manner in which language brings man to speak. In the essay *Hölderlin and the Essence of Poetry* Heidegger refers to this manner and accordingly the essential nature of man, by saying "we - mankind- are a conversation."[116]

Human conversation consists in the first instance of acts of speaking with others about something. This speaking is complemented by, and presupposed by, acts of listening. By speaking and listening, a particular conversation brings human beings together. In this way conversation points to the essential function of language as that which unifies humans in their understanding of that which is manifest in conversation. Hence the nature of the human encounter with language is to be discerned

---

[112]PLT. 210.

[113]WL. 112.

[114]WL. 112.

[115]WL. 112.

[116]M. HEIDEGGER, "Hölderlin And The Essence of Poetry" in W. BROCK, *Existence and Being*, 227. K. Harries sheds much light into Heidegger's interpretations of Hölderlin and language, see, K. HARRIES "Heidegger and Hölderlin: The Limits of Language" *The Personalist* 44 (1963) 5-23.

only insofar as man undergoes an experience with language.

Heidegger in *The way to Language* clearly defines his task as that of 'experiencing the unbinding bond within the web of language' [117] Companion essays to that one, *The Word* and *The Nature of language*, similarly insist on undergoing an experience with language. In the essay *Language* Heidegger indicates the basis for the possibility of experiencing language authentically: "In any case language belongs to the closest neighbourhood of man's Being. We encounter language everywhere."[118] Language as language is always at least as close to the Being of man as anything else. This point is assured by the fact that language is the foundation of human Being.[119] In the essay, *The Nature of Language*, the transformation of man's relation to the essence of language is named as 'undergoing an experience with language' (*mit der Sprache eine Erfahrung machen*):

> To undergo an experience with language ... means to let ourselves be properly concerned by the claim of language by entering into and submitting to it.[120]

Heidegger maintains in the same text that 'an experience we undergo with language will touch the innermost nexus of our existence' because the proper abode of man's existence is in language. Thus, speaking about the essential nature of language in relation to the Being of man, Heidegger considers language as a gift to man, a gift which is determinative of the nature of man[121]

---

[117]WL.113.

[118]PLT. 189. For an evaluative study of "The Word" (*Das Wort*), see, P. McCORMICK "Heidegger's Meditation on the Word," *Philosophical Studies* 18 (1969) 76-79. Bernasconi rightly points out the development of two views on language in "The Word." See, R. BERNASCONI, *The Question of Language in Heidegger's History of Being* (Atlantic Highlands, NJ: Humanities Press 1985), 49f.

[119]WL. 112.

[120]WL. 57.

[121]HEP. 280.

in that man alone amongst all beings speaks. This gift of language provides an abode to man wherein man dwells.

To undergo an experience with language requires that we break out not only of the experience of language as an object but also of the experience of language as something proper to subjectivity. In other words, to undergo an experience with language is not to engage in reflection on language in that sense according to which language would, in advance, have been taken as subjective activity accessible essentially only to the higher-level subjective activity of reflection. Hence, to gain an experience with language does not mean that we should engage in experiments with language but that, 'to let ourselves be properly concerned by the claim of language by entering into and submitting to it while reflecting on our abode in language.[122] In other words, we should become fully aware of something that immediately concerns our own Being. Heidegger refers expressly to the fact that the issue is not to gather knowledge about language in the sense of metalanguage and metalinguistics. "Metalinguistics is the metaphysics of the universal technification of all languages into the only functioning interplanetary instrument of information."[123]

In opposition to the scientific and philosophical knowledge of language, Heidegger proposes "to gain an experience with language." Language has the special characteristic that we live in it, are familiar with it, and deal with it without catching sight of it. We continuously heed what becomes accessible to us through language and thereby overlook language itself. In order to get out of this position Heidegger again appeals to the poet, not merely because he has a privileged relationship to language but because he brings this relationship up for discussion. Whereas Hölderlin poetizes the essence of the poet, Stefan George poetizes the

---

[122]WL. 57.

[123]WL. 58.

relationship with language, our experience with language. This is why at the centre of these explanations Heidegger interprets George's poem "A Winter Evening."[124]

Heidegger takes up George's poem "A Winter Evening" as an occasion to elucidate his own view on the relationship between Being (taken as simultaneously including Fourfold, World and *Logos*) and language (taken as the original stillness in which all things find rest). He wants to establish that the "Being" (*Wesen*) of original language needs man's language and, by reason of this need, ap-propriates to man what is proper to himself, in order to ap-propriate man to itself in the process of its own "coming to pass." This means as Richardson points out that man's language must be understood as his hailing response to the hail of aboriginal language as it abidingly comes to pass (*west*).[125] When a man speaks, he takes up a language that is already constituted; in his speech, he listens to what this language has to say.

To say something means to point something out, to show it. The very essence of language, then, is to be found in the fact that language says something, shows something, lets something appear. Man must listen to language in such a way that he lets its saying speak itself out to him. In his own speaking, which essentially implies his listening to language, man must "say after" what he has heard before. But if man's own speaking essentially implies his listening to language, then his speaking can come about only insofar as his being was already open to language's saying. Man hears language's saying only to the degree that he belongs to the domain which language discloses to him. Only to those who "belong" to language does language grant the

---

[124]In 1950, Heidegger delivered the lecture entitled *Die Sprache* (Language) in which he made use of George Trakl's poem *Ein Winterabend* (A Winter Evening) to get at what he regards as the essence of language.
[125]W.J. RICHARDSON, *Phenomenology to Thought*, 577-578.

possibility of listening to it and, consequently, the possibility of speaking as well.[126]

The relationship of language and Being makes it evident that this relationship is dependent upon the thoughtful dwelling of man within the truth of Being. Since this dwelling of man finds its expression in language, it becomes evident that for Heidegger there is a special relationship between language and man. On the one hand, language is used by man and is commonly considered as the vehicle of the transmission of human significations; on the other hand, man dwells in language, within a realm of a totality of meaning. For Heidegger this realm of the totality of meaning is directly referable to the truth of Being.

Human speech says Heidegger "as the speech of mortals rests in its relation to the speaking of language."[127] "Language requires human speaking and is at the same time not the mere product of our speech activity." [128] The speaking of language, as it comes to light, is a saying that shows by calling each called-to into its own-most self. It is to such speaking that human speaking must dedicate itself in order to be authentic human speaking. It is this emergence of the phenomenon of being called into its own-ness, that precludes language from being a mere product of human activity.

Heidegger asserts that "Man speaks in that he responds to language. This responding is a hearing. It hears because it listens to the command of stillness"[129] What Heidegger is saying here, is that it is only in so far as human speaking is dedicated to the speaking of language that human speaking is authentic, or that human being comes into its own-most essence as the living being

---

[126]PLT. 189-193; WL. 47f.

[127]PLT. 208.

[128]WL. 125.

[129]PLT. 210.

who speaks. Human speaking belongs to the speaking of language,[130] to the tolling of the bell of silence, in so far as human speaking is first of all a listening to, a hearkening to the speaking of language. Only after so listening can human speaking speak in its own way.

Heidegger writes: "Man behaves as though *he* were the shaper and master of language, while in fact *it* remains the master of man."[131] This says: the transformation required for the restoration of language to its element cannot take the form of a reshaping or reforming of language, as though language were no more than a matter in need of being appropriately informed through the activity of a subject. What is called for is not a transformation of language as though it were an object over against man, to be worked over by man. What is called for is rather, a transformation of man's 'relation to the essence of language'[132] - a transformation which would shatter all pretense to objectification. As Fay opines if language is viewed as the abode in which man dwells, the only habitation in which he can abide as man, then it becomes clear that a threat to language is a threat to man himself.[133] To destroy man's language then, is to destroy the only element in which man can live as man.

10. *DIE SAGE*: ESSENCE OF LANGUAGE

In IM Heidegger states that his aim is "an essential clarification of the essence of being in respect to its essential involvement with

---

[130]"Belong" in German is *gehören* whose root is *hören*, i.e., to hear, to listen to. This word-play opens us to a more significant understanding of the relation between human speaking and the speaking of language.

[131]M. HEIDEGGER, *Vorträge und Aufsätze* (Pfullingen: Neske 1954), 146.

[132]WL. 135.

[133]T.A. FAY, "Heidegger on Logic and Language: some *Aporiai*" *Tijdschrift voor Philosophie* 52(1990) 436-7.

the essence of language."[134] Inasmuch as Heidegger's aim has been to work out the problem of Being concretely, his working-out of the auxiliary problematic of the nature of language is likewise found to search for a concrete understanding of the nature of language. The determination of the essence of language, the very inquiry into it, are regulated at all times by the prevailing preconception about the essence of being and about essence itself. But essence and being express themselves in language.[135] Language is the "house of Being"- an abode given by Being to man, within which abode man is to dwell thoughtfully.[136] Heidegger elucidates the point thus:

> ...language is the house of Being, which is propriated by Being and pervaded by Being. And so it is proper to think the essence of language from its correspondence to Being and indeed as this correspondence, that is, as the home of man's essence. ...language is the house of being in which man ek-sists by dwelling, in that he belongs to the truth of Being, guarding it.[137]

The essay *Hölderlin and the Essence of Poetry* depicts language as the supreme event of human existence which is determinative of the nature of man[138] In IM we see that in language and words things first come into Being and are.[139] Further, in the notion of Being as appropriation Heidegger makes evident that the problem of the nature of language is necessarily connected with the problematic of the Being-question. For man and Being belong together in the abode of the primal speaking of language. Taking the perspective of these passages, it is evident that the later Heidegger does not consider language to be simply a being among beings, because he considers language to hold sway at once over

---

[134]IM. 54.

[135]IM. 53.

[136]LH. 217.

[137]LH. 236-237.

[138]HEP. 280.

[139]IM. 13.

the existence of man and over the presence of beings. There is therefore a very special relationship between language and Being. Entering into the problem of the nature of this relationship Heidegger cites a passage from the Greek tragic drama *Antigone*: "simultaneously with man's departure into Being he finds himself in the word, in language."[140]

In IM the word *logos* understood as "the gathering that makes manifest," is determinative of the essence of man and thereby determines the essence of language. In and through language the Being of beings is made manifest: "In accordance with its ... essence, being-human is *logos*, the gathering and apprehending of the essent of them all, in whom ... through acts of power, the overpowering is made manifest and made to stand.[141] Thus language understood in terms of this interpretation of *logos* determines the existence of man as he stands in the clearing of Being. However any attempt to ground language in Heidegger's view is bound to fall into absurdity because it moves within language and is constrained to examine language from within language itself. So he states mysteriously and tautologically: "Language is language" [142]and concedes that the essential nature of language remains a mystery. The fact that man can ask questions about the nature of language requires that language be

---

[140]IM. 171. Heidegger translates the Greek word *logos* as 'language' which is one of the key words of Heidegger's retrieval of the thought of the Greek pre-Socratics. As T.A. Fay observes, by the time of Heraclitus' interpretation which is presented in the essay *Logos* published in 1951, the union between *logos* as essence of language and Being has been so successfully established, at least in Heidegger's mind, that *logos* and Being are seen as inseparably bound together; *logos* is in fact another name for Being, Being as gathered unity lying forth in openness. T.A. FAY, "Heidegger on Logic and Language: some *Aporiai*," 434.

[141]IM. 143.

[142]PLT. 191.

already granted to him. Thus, authentic thought about language does not take the form of an inquiry, but is rather a listening to the grant of language.[143]

In fact language withholds its nature from words and the prospect of the thinking that labours to answer to the nature of language is still veiled.[144] However, by reflecting upon the grant of language man can arrive at his abode in language. "To reflect on language means ... to reach the speaking of language in such a way that this speaking takes place as that which grants an abode for the being of mortals."[145] Thus, man will come into the essential dimension of language by way of reflection on its speaking. To reflect upon the grant of the speaking of language is the most basic positive response to the gift of language, a response which is of the order of thankfulness. Original thanking is the echo of the favour of Being itself and thanking as the speechless response to the favour of Being is the source of human language.[146] In the essay *The nature of Language* Heidegger indicates the inadequacy of speaking about language:

> ... Language is always ahead of us. Our speaking merely follows language constantly. Thus we are continually lagging behind what we first ought to have overtaken and taken up in order to speak about it. Accordingly, when we speak of language we remain entangled in a speaking that is persistently inadequate.[147]

'That which must remain wholly unspoken is held back in the unsaid, abides in concealment as unshowable, is mystery.'[148] And this is the fate of the essence of language which remains hidden in mystery for us. Heidegger in *The way to language* makes one

---

[143]WL. 71.

[144]WL. 8.

[145]PLT. 192.

[146]WM. 358.

[147]WL. 75.

[148]WL. 122.

more attempt and immediately concedes that it has failed according to the standard he has already recognized: 'with regard to the manifold ties of saying we shall call the essence of language in its totality die. *Sage*- and admit that even now we have not caught sight of what unifies those ties'[149] According to R. Bernasconi, *Sage* is Heidegger's word for that originary language through which the destiny of Being happens and to that extent goes beyond what in BT was named as *Rede*.[150] There is no word for the essence of language according to the metaphysical way of naming; indeed the attempt to name this essence is itself only a metaphysical ambition if it means to crack open and divulge the mystery of the unsaid.

To bring the essence of language to language, now comes to mean to hear the essence of language in every word, or one might say to enter into the essence of language to which in a sense we already belong. The essential being of language is the *Sage* as that which shows[151] (*Des Wesende der Sprache ist die Sage als die Zeige*). The showing of saying takes place when language breaks in to speech. It is language itself which reaches into the regions of presencing and lets what-comes-to-presence appear and disappear. Human saying is only a *Nachsprechen*, literally a 'saying after,' a reiteration.[152] Heidegger continues thus: Before man speaks he must first let himself be claimed again by Being, taking the risk that under this claim he will seldom have much to say. Only thus will the pricelessness of its essence be once more bestowed upon the word, and upon man as a home for his dwelling in the truth of Being.[153] And "If we consult language

[149]WL. 122-123.

[150]R. BERNASCONI "The Transformation of Language at Another Beginning" in C. MACANN, *Martin Heidegger,* 179.

[151]WL. 123.

[152]WL. 124-125.

[153]LH. 223.

about its essence, then language itself must already be addressed to or spoken-to us. If we inquire about the essence, namely, of language, so must also the meaning of essence (*Wesen*) be already spoken-to us."[154]

11. *SAGEN*: THE SHOWING OF SAYING

Now the essence of language, for Heidegger, is Saying (*die Sage*) as showing (*als die Zeige*)- a showing that raises up a totality of relations or a world within which any given thing has a meaning. Heidegger says this quite explicitly. "To say ...means to show: to make appear, to set free, that is, to offer and extend what we call world, lighting and hiding proffer of world is the persisting nature of Saying."[155] To show then means to allow something to appear, to set it free to its world. Thus it appears that something comes to presence on its own terms, and not on those of a subject, when it appears in its world. What is shown and handed down by language's Saying is, in the final analysis, the world. The mode of Being characteristic of this Saying is to render the world free in a revealing and, at the same time, concealing way. The Saying of language lets what is present in the direction of a permanently being present just as it fetters what is absent in its permanently being absent. The saying thus "joints"

---

[154]WL. 71.

[155]WL. 93. Because Heidegger finds an affinity between *Sagen* and *Zeigen* (to show-forth, to let-appear-in-the-open) Richardson, translates *Sagen* as "utter" and the word "utter" derives from the comparative form of the Anglo-Saxon *ut*, meaning "out," hence may be taken to mean "to give or bring out," sc. in the Open. (W.J. RICHARDSON, "Heidegger and the Origin of Language," *International Philosophical Quarterly* 2 (1962) 41. As prof. W. Marx points *Sage* for Heidegger is, the non-human, silent Saying, integrally belonging to *alētheia*, emanating from *lēthē* (concealment [*Geheimnis*]), and permeating all Saying. W. MARX "The World in Another Beginning: Poetic Dwelling and the Role of the Poet" in J.J. KOCKELMANS, *Heidegger and Language,* 250, note.59.

the openness of the "clearing" for which each appearing looks and from which each concealing flees. The Saying is the gathering together (*logos*) of a manifold pointing that joints together all that appears and lets what was shown sojourn everywhere by itself. [156]

Heidegger confirms that "we shall call the being of language in its totality "Saying."[157] Language is not here defined in terms of something other than itself, i.e., as human activity, in terms of phonetics and conventions, rather it comes forth as it itself is, i.e., as a Saying that shows, that manifests. Language shows itself at first as our way of speaking and involves a manifold of elements. We find thus: The Speakers, That where of they speak, What is spoken and The unspoken. 'The spoken stems in manifold from the unspoken, whether this be something not yet spoken or something which must remain unspoken in the sense that it is withheld from speaking.'[158] The essential Saying of language therefore is the primal Saying which shows, or manifests, the Being of beings in presence. The primal Saying of language which is prior to all human speech is a monologue.[159] This monologue is characterized as "stillness (*Stillen*)"[160] Yet this stillness, as Ambrosio suggests, is not stagnant immobility, the absence of vibrant motion.[161] It is the harmony of identity, of mutuality and appropriation so perfectly attuned that it rings out in a tone so mellow that it is no sound but a song, an impulse, complete in itself, yet open in that it calls all things into their being.

Men are admitted into primal Saying and belong within it. Saying grants to a man the ability to hear the speaking of his

---

[156]WL. 126.

[157]WL. 123.

[158]WL. 120.

[159]WL. 134.

[160]PLT. 207.

[161]F.J. AMBROSIO, "Dawn and Dusk: Gadamer and Heidegger on Truth" *Man and World* 19 (1986) 29.

language and to attain the ability to speak. The essence of language is present in Saying as the source of such grant.[162] Thus, the essence of language grants to man the gift of speech, which gift determines his human nature. At the same time language needs and uses the speaking of humans in order to bring the primal Saying to expression. In the essay *Language* Heidegger brings out this point:

> ... the very nature, the presencing, of language needs and uses the speaking of mortals in order to sound as the peal of stillness for the hearing of mortals. Only as men belong within the peal of stillness are mortals able to speak in their own way in sounds.[163]

Language grants the gift of speech to man and hence, human speech is bound to the essence of language from within language itself. Further language commands human speech by bidding the human response of listening to its primal Saying.[164] Inasmuch as the essence of language dominates human speech in the two-fold manner of a grant and a command, it can be properly said of language that it owns man in the sense that language makes man appropriate to its own Saying.

The Saying is carefully differentiated from human speaking, even though human speaking is also included in the layout. Man belonging to Saying (*der Sage gehörend*) -listens to it and its word and brings what it hears "correspondingly" into human-sounding words. Mortal speech is thus characterized as a co-respondence (*ent-sprechen*) or a response (*Nachsagen*).[165] Hence we must listen to a language in such a way that we let its saying "speak itself out" to us. This fundamental willingness to let a language speak itself out to us precedes all hearing and listening. In our own speaking, which essentially implies our listening to a

---

[162]WL. 124.

[163]PLT. 208.

[164]PLT. 208-209.

[165]WL. 134.

language, we "say after" what we heard before. We let the "soundless voice" of language come over us and evoke the sound that is "locked up in us."[166]

We do not speak a language as much as we speak from a language. We are able to do this only because we have first listened to that language before we have started to speak that way. What we hear and listen to in this case is the Saying of a language. Our own speaking, which essentially implies our listening to a language, can allow language as Saying to speak itself out to us only in so far as our own Being was already open to such Saying. We hear the speaking of a language only to the extent that we belong within the domain which it discloses for us in speaking. Only to those who belong to a language does the language grant the possibility of listening to it and, consequently also, the possibility of speaking. This granting, which goes on continuously in each concrete act of man's speaking, is what lets us attain the possibility of speaking. In the granting of primordial Saying the essence of language is to be found.

However, in our speaking, language itself remains largely concealed, it withholds itself, holds itself in reserve, withdraws. Heidegger says thus: "Only because in everyday speaking language does not bring itself to language but holds back, are we able simply to go ahead and speak a language, and so to deal with something and negotiate something by speaking."[167] But this self-withholding is nothing accidental nor is it something which occurs only in the sphere of unreflective everyday speaking. It belongs to the essence of language itself that it comes to presence only in such fashion as also to hold itself decisively in reserve; that is, language never comes fully to language: "For this mystery the word is lacking, i.e., that saying which would make it possible to

---

[166]J. J. KOCKELMANS, "Language, Meaning, and Ek-sistence" in J. J. KOCKELMANS, *On Heidegger and Language*, 24-25.

[167]WL. 59.

bring the essence of language- to language."[168] Language intrinsically withdraws, withholds itself, and "this withholding belongs to the essence of language."[169]

Language is a response to the Saying of aboriginal language (*logos*). What does language summon into presence? What is summoned in the coming to pass of language as *logos* is the correlation of Being and beings, and both are summoned toward the unifying scission of the difference that prevails between them. The original dif-ference calls out to human talk to let world and thing be in their intimacy and difference:

> *Language speaks as the peal of stillness.* Stillness stills by the carrying out of world and thing in their presence. The carrying out [*Austrag*] of world and thing in the manner of stilling is the appropriative taking place [*Ereignis*] of the dif-ference [*Unter-Schied*]. Language, the peal of stillness, is, inasmuch as the dif-ference takes place.[170]

Language is nothing human; on the contrary, man is something linguistic, given to speech. Man belongs to the essence of language by being required by language for language to come to words: "the coming-to-presence [*Wesen*] of language as the peal of stillness needs and uses [*braucht*] the speaking of mortals in order to sound as the peal of stillness for the hearing of mortals. Only insofar as men belong within the peal of stillness are mortals able to speaking their own way in sounds."[171] In speaking man responses to a call which is addressed to him by the dif-ference.

Language speaks by bidding men; men speak by heeding the

---

[168]WL. 154.

[169]WL. 81.

[170]PLT. 207.

[171]PLT. 208. In W. BIEMEL'S view the relationship between the speaker and language recalls the relationship between Dasein and Being that Heidegger mentions earlier, when he says that Dasein can be only by the grace of Being, but on the other hand Being is in need of Dasein. Language is in need of man's speech and is nevertheless not the mere product of his speech activity.

bidding. The dif-ference is by its nature for Heidegger a linguistic event, in the sense that it is the happening of language. Beings appear in the opening which the dif-ference opens up. But this opening is cleared by language. "In the bidding that calls thing and world, what is really called is: the dif-ference."[172] Language, originally seen, is this scission, as the coming to pass of the ontological dif-ference out of original *logos*. The ontological dif-ference is a scission (*Schied*) between (*Unter*) Being and beings that refers them to each other by the very fact that it cleaves them in two.[173]

The poetic speech, according to Caputo, lets the difference between world and thing come into presence. Hence, the dif-ference itself requires language.[174] The difference between world and thing emerges from the dif-ference itself which is essentially linguistic. The dif-ference is at once both what calls thing and world and that into which they are called. It is that which calls inasmuch as it is the silent source - here the "stillness"- from which thing and world emerge. Then, when this call is taken up into human language, when man hears this address, thing and world and so the intimacy and dif-ference in which they subsist come into words. The dif-ference stirs human talk to name world and thing. But the original call of the dif-ference, since it precedes human talk, is itself silent.

The original and silent call of the dif-ference, which Heidegger calls the "peal of stillness,"is the speaking not of man but of language itself. The difference is at the same time unity and "two-ness", hailing and hailed, differentiating and differentiated, and the scission as such is nothing but the tension and mutual adhesion of unity and duality.[175] Authenticity in the use of language is

---

[172]PLT. 203.

[173]W.J. RICHARDSON, *Phenomenology to Thought*, 578-579.

[174]J.D. CAPUTO, *Heidegger and Aquinas*,161.

[175]PLT. 207.

achieved at the moment of man's free response to the hail addressed to him when the scission takes place. It is man's task to respond to the hail addressed to him out of the need of *logos* for a *Da* in order that the differentiating may give rise (issue) to the differentiated. The adequate response to this hail is the coming to pass of man's language in complete authenticity.[176]

In short, for Heidegger language is not adequately and appropriately grasped when it is construed merely in anthropological and instrumental terms. For him language has a unique showing power that goes deeper than that. When language unfolds essentially, it allows things to show themselves and be manifest. The root unfolding of language occurs as 'a way-making' (*be-wëgen*) so that things may appear and show themselves. The word way and what it indicates requires that we think of Being and language, not as two separate and independent entities, but as always connected and in accord. They are distinct from each other but are not separate and independent of each other. Hence to think of language we must think of its words' showing power, which is always a showing power that shows things in their being. When we state something in words, we always show something in its being. In short, every statement in language is stated in the horizon of the root unfolding of Being; and being appears within the horizon of the root unfolding of language.

12. LANGUAGE, THOUGHT AND POETRY

Language, Heidegger argues, is pure spokenness. Now he flatly declares that whatever is purely spoken is a poem.[177]

---

[176]W.J. RICHARDSON, *Phenomenology to Thought*, 578-581.

[177]PLT.194. Heidegger acknowledges the superiority of poetical discourse. A reading of his later texts reveals that he was strongly influenced by the Greek poet Parmenides and the German poet Hölderlin and to a lesser but important extent by Sophocles and the German poets George and Trakl.

Heidegger conceives of the essence of poetry as "the establishing of Being by means of the word."[178] In *Hölderlin and the Essence of Poetry* Heidegger indicates that the role of poetry is to bring originally into the openness of the clearing of Being that which is subsequently dealt with in everyday language.[179]The poetry takes its place within the realm of the speaking of language by naming that of which it speaks. "... it brings the presence of what was previously uncalled into a nearness."[180] Hence the poem as it names brings that which it names into presence. Thus it brings the presence of what was previously uncalled into the nearness of presence, i.e., into Being.

According to Heidegger language is the primordial poetry through which a people speaks Being. Conversely, the great poetry by which a people enters into being initiates the moulding of its language. The Greeks created and experienced this poetry through Homer. Language was manifested to their being-there as a departure into Being, as a configuration disclosing the essent.[181] Heidegger views the poet as he who lets things be present in the higher sense, namely as things called forth which reflect the world-foursome and the four world-poles of the interplay. The fact that the poet calls things forth suggests that he has a special relation to language. It is precisely the poet who does not treat language as mere instrument; part of his "art" is to allow the beauty and movement (*Bewegtheit*) of language to unfold. He is sensitive to the subtle undertones of language, those tones which the everyday ear fails to hear.

> The speech of the poet is the interception of these signs (signs to us from antiquity, the language of the gods) in order to pass them on to his own people. This intercepting is an act of giving; for "in the first

---

[178]HEP. 261.

[179]HEP. 283.

[180]PLT. 198.

[181]IM. 171-172.

> signs" the poet catches sight already of the completed message and in his word boldly presents what he has glimpsed, so to tell in advance of the not-yet-fulfilled.[182]

Above all the poet is listener; were this not so, he could not be sensitive to the lyrical quality of language.[183] All poetry is not of equal quality; only thoughtful poetry reflects the requisite disclosure of language. Such thoughtful poetry or poetic thought occurs only at the point of proximity between poetry and language: Saga, *Sage*. Hence the essence of poetry must be understood through the essence of language:

> First of all it appeared that the field of action of poetry is language. Hence the essence of poetry must be understood through the essence of language. Afterwards it became clear that poetry is the inaugural naming of being and of the essence of all things- not just any speech, but that which for the first time brings into the open all that which we then discuss and deal with in everyday language. Hence poetry never takes language as its raw material ready to hand, rather it is poetry which first makes language possible. Poetry is the primitive language of a historical people. Therefore, in just the reverse manner, the essence of language must be understood through the essence of poetry.[184]

Here Heidegger refers to language as "primitive language." True poetry, he says, is never merely a higher mode of everyday language. Every day language is described as forgotten and used up poetry from which scarcely any calling forth can be heard.[185]

---

[182]HEP. 287. As L.M. VAIL observes Heidegger claims that poetry provides pure examples of language because in poetry language operates as an unfolding power rather than as a mere instrument. The poet especially cannot treat language as a mere instrument; one of the features of good poetry is that it brings the richness and beauty of language to light. For the poet language is his element as well as his tool; for this reason poets and poetry are given a great deal of attention by Heidegger. L.M. VAIL, *Heidegger and the Ontological Difference*, 161.

[183]WL. 188.

[184]HEP. 283-284.

[185]PLT. 208.

In the essay *The Origin of the Work of Art* Heidegger speaks of all art as the letting happen of the advent of the truth of what is, which for him is essentially poetry.[186] But since language is the happening in which beings first disclose themselves to man each time as beings, language itself is poetry in the essential sense. Language alone brings what is, as something that is, into the open for the first time. By naming beings for the first time, language brings them to word and appearance originally. Only this naming nominates beings to their Being from out of their Being. Where there is no language, there is no openness of what is, no world is set up nor is the earth set forth, and the truth of beings does not occur.[187]

The spirit of poetry is essentially superior to the spirit that prevails in all mere science.[188] We note here a pre-eminence given to poetry with respect to science. Science deals with things theoretically in their present-at-hand kind of Being and with a language based on the assertive modality adequate to articulate this kind of intelligibility. By implication, then, the poet's superiority lies in the fact that poetical discourse is more primordial in that it deals with things in their ready-to-hand kind of Being within the pre-scientific world of Dasein.

As Munson points out Heidegger visualizes the task of poet and philosopher as essentially the same: the communication of an original experience. The poet questions "the world" and listens to "the world." The philosopher on the other hand is pledged to a similar endeavour with *Sein.* He seeks a pre-ontological understanding of Being which plunges him into the structures of Dasein itself to understand its possibility of deploying a world, of

---

[186]PLT. 72. Essential poetry includes poesy, or what we ordinarily refer to as poetry, indeed in a privileged way, but extends beyond it to all genuine forms of art.

[187]PLT. 73.

[188]IM. 26.

"letting things come to be." Like the poet, the philosopher questions and listens to "Being."[189] "The poet must renounce having words under his control as the portraying names for posited being."[190]

The renunciation of the poet is twofold. J. Sallis explains it thus: in the first place, the poet renounces that representation according to which things are constituted in their essential being independently of language and of everything linguistic. He renounces that view according to which beings would always already have come to presence in advance of naming which poet would then bring to them; he renounces the belief that to name is merely to portray what is already fully established in its presence. Positively turned, the poet recognizes that words "bething" things - that is, first allow things to come forth into presence, to be. Secondly, the poet renounces having words under control, under his mastery, as though they were mere instruments to be shaped and used at his pleasure. The poet renounces his claim to be master of language. Positively turned, he grants language its intrinsic elusiveness; he lets it openly escape, lets it withdraw into its appropriate self-reserving concealment. He lets language announce its mastery.[191]

Heidegger writes that a dialogue between thinking and poetry is possible "because a distinctive, though in each case different, relation to language is proper to them both."[192]He says further: "The dialogue of thinking with poetry is directed towards calling forth the essence of language, so that mortals may learn again to

---

[189]T.N. MUNSON, "Heidegger's Recent Thought on Language," *Philosophy and Phenomenological Research* 21/3 (1961) 362.

[190]WL. 147.

[191]J. SALLIS "Towards the Showing of Language," *Southwestern Journal of Philosophy* 4/3 (1973) 78.

[192]WL. 161.

dwell within language."[193] The dialogue of thinking with poetry - a dialogue between thinking and poetry which, however, is presided over by thinking and not by poetry- is charged with calling forth the essence of language for thinking, even if not yet in the way that is entirely appropriate for thinking. Poetry offers something to thinking, not when thinking merely appeals to poetry for attestation or elaboration of something which thinking could itself establish, but rather when thinking attends to poetry in such a way as to be able to hear sounded in the poetic word that restoration of language to its element by which, in turn, thinking is first itself enabled to come into its own element, to maintain itself appropriately in its distinctive relation to language.

Thinking, traditionally the function of understanding, apprehends and gathers into unity (as *noein* and *legein*), conveys correspondence in the element of language: *das Denken als Entsprechen steht im Dienst der Sprache*, "thinking as corresponding stands in the service of language."[194] Man, as thinking, does not manipulate words as instruments with the purpose of communicating some message, but always already 'dwells' in language, 'the house of Being and shelter of the essence of man.' And yet Heidegger maintains that thought 'acts': "Thought is attentive to the clearing of Being whenever it inserts its Saying into language. ...In this way thought is a doing (*Tun*)."[195] Thinking acts to the extent that it "brings Being to language." (*Zur Sprache bringen*). "Thinking gathers language into a simple saying. Language is then the language of Being, like the clouds are the clouds of the heavens."[196]

The true thinker is characterized by his closeness to the essence of language. This is the reason for the Heideggerian

---

[193]WL. 161.

[194]WP. 44.

[195]WM. 239.

[196]LH. 265.

meditation on the relationship between the poet and the thinker. Both the poet and the thinker are the guardians of language as the 'house of Being.'[197] Dasein as the openness of Being is disclosure which is the fundamental truth for Heidegger. Truth is related to thinking. According to Heidegger, true thinking is primary thinking. Such thinking does not stand in man's disposition and is never his instrument. It is a response of man to Being rather than his direct saying.

Primary thinking occurs as an echo "to the Word of the soundless voice of Being."[198] It is a thinking belonging to Being.' The German words '*zuhören*' (listen) and '*zugehören*' (belong) sound alike and in some forms are identical. Heidegger uses these words in both aspects to express man's subordination to Being in his thinking. By listening to Being we belong to it. By belonging to it we are what we are. By responding to Being we are led to our self. By thinking we preserve or guard the truth (revelation or disclosure) of Being in our thoughts or words. Our ability to think is, in fact, our ability to respond to the Being. Therefore, and only in this way, can our thought or words be true.

13. LANGUAGE: EVENTING OF BEING

Heidegger calls presence or disclosedness by the name 'Being.' "...the uncovering of the undisclosed into disclosedness is the becoming-present itself of things present. We call this the Being of being."[199] If language receives its essential form from the disclosedness of the "lying-before-together," then language essences, or comes to presence itself, as making present (showing), and thus itself comes to presence as given over to presence, to disclosedness, to Being. In so far as language

---

[197]LH. 217.

[198]WM. 358.

[199]M. HEIDEGGER, *Early Greek Thinking*, tr., D.F. KRELL & F.A. CAPUZZI (New York: Harper San Francisco 1975), 76.

receives its essence from disclosedness, mystery (*geheimnis*) lies at the heart of language. Heidegger says so explicitly:

> In connection with the question of the essence of language, the question of its origin has arisen time and time again ... The origin of language remains a mystery; ... The origin of language is in essence mysterious. And this means that language can only have arisen from the overpowering, the strange and terrible through man's departure into Being.[200]

As we have already discussed instead of explaining language in terms of anything else Heidegger prefers 'to let language be experienced as language.'[201] "To discuss language, says Heidegger, means not so much to bring it, but rather us, to the locus of its essence: gathering in "*das Ereignis*."[202] Here Heidegger seems to be saying that the locus of the essence of language is equivalent to or it is in "*das Ereignis*"(Event of Appropriation). In *The way to language* Heidegger can be found ascribing the origin of the language to Appropriation (*Ereignis).* He says: "All human language is appropriated in Saying and as such is in the strict sense of the word true language- though its nearness to appropriation may vary by various standards."[203] It seems that the fundamental locus of language has changed and emerges in the closest neighbourhood to Being. Thus Heidegger's problematic of Being brings him to understand Being concretely as the "Event of Appropriation (*Ereignis*)." The understanding of Being as Appropriation arises out of his meditation of the Greek pre-Socratic usage of the word *logos*. In his retrieval of this key Greek word Heidegger finds that it does not merely signify 'language' or 'reason,' but indicates the primal gathering of beings in Being. Accordingly in the later period language, as

---

[200]IM. 171.

[201]WL. 119.

[202]WL. 10.

[203]WL. 133.

founded in *logos* transcends the individual human being, and is in fact the gathering of the manifestness of beings within the enduring presence of Being.

In the essay *Hölderlin and the Essence of Poetry* Heidegger maintains that within language there is manifestness of Being as presence. "... Language is not a mere tool, one of the many which man possesses; on the contrary it is only language that affords the very possibility of standing in the openness of the existent."[204] Further it is language which guarantees for man the openness of the clearing of Being and the presence of beings in Being:

> ... language alone brings what is, as something that is, into the Open for the first time. Where there is no language, as in the beings of stone, plant, and animal, there is also no openness of what is, and consequently no openness of either that which is not and of the empty.[205]

In the same essay Heidegger continues: "Projective saying is saying which, in preparing the sayable, simultaneously brings the unsayable as such into a world."[206] Heidegger's understanding of primal Saying seems to be bound up with his understanding of Being as appropriation.[207] In ID appropriation is "the belonging together of man and Being in the manner of mutual challenge" in which "man is delivered over to the ownership of Being and Being is appropriate to the essence of man."[208] In the context of his understanding of appropriation Heidegger finds the nature of language as relatedness. He states: "Language ... is the keeper of being present, in that its coming to light remains entrusted to the appropriating show of Saying. Language ... as Saying is the mode

---

[204]HEP. 275-276.

[205]PLT. 73.

[206]PLT. 74.

[207]WL. 127.

[208]ID. 36.

of appropriation."[209] As he continues in this passage, he indicates that language as the mode of appropriation is 'that appropriating, holding, self-retaining (which) is the relation of all relations.' Human speaking, inasmuch as it is a response to the call of Being likewise 'remains forever relational.'

Appropriation is the sounding of the primal word which makes human nature 'appropriate for that which avows itself from everywhere to man in Saying ...'[210] Accordingly man is made appropriate for the Showing of primal Saying so that he may encounter primal Saying and respond to it. In order to be engaged in authentic speech, a man must pursue the way which is laid open by the appropriation of primal Saying. Insofar as a man pursues this way, he undergoes a transformation which is determined by the destiny of his belonging together with Being in appropriation. Therefore, there is an essential connection here between *Ereignis* and language and the abode of mortals. Heidegger says: "*Ereignis* grants to mortals their abode within their nature and that they may be capable of being those who speak.[211] Heidegger adds further: "... thinking must first overcome the habit of yielding to the view that we are thinking here of 'Being' as *Ereignis*. But Ereignis is different in nature, because it is richer than any conceivable definition of Being. Being however, in respect of its essential origin can be thought of in terms of *Ereignis*."[212] *Ereignis* as the very centerline of his endeavours, Heidegger asserts that 'it can be translated with as little success as the Greek keyword *logos* and the Chinese *Tao*.[213] In fact, he later specifies the *Ereignis* as the one and only, unique and simple subject matter of thought, inaccessible in its simplicity, approachable only through a

---

[209]WL. 135.

[210]WL. 129.

[211]WL. 128.

[212]WL. 129.

[213]ID. 29.

manifold thought, and *ipso facto* through a manifold language.[214]

According to W. Biemel the manifold possibilities of showing refer to the saying as that which shows, and this, in turn, refers to the *Ereignis*. He continues 'it may be appropriate here to remember that we are not permitted to hypostatize the *Ereignis* as a power which is beyond everything and which holds sway over being; we must rather try to grasp the *Ereignis* as that which governs in language and which we run into in our questioning back concerning language's pointing. Genuine speech is for Heidegger a cor-responding to the Saying and to the appropriating event. In genuine speech nothing takes place but a manifestation of the appropriating e-vent, which remains hidden, however, for the one who speaks. That is why, according to Heidegger the thinking experience of the essence of language is nothing but the freeing of the movement that leads from the appropriating e-vent to man's speech. Language is able to grant the clearing because in its very essence language is a granting and an appropriating e-vent.'[215]

14. CONCLUSION

In the present chapter we have undertaken a study of Heidegger's understanding of the nature of language in relation to his understanding of Being over each of the two distinct periods in his career - the period of BT and his later period which follows the publication of BT. In BT Heidegger presents language as a constitutive element of the Being of Dasein, where the term Dasein indicates the existence of the individual human being. In the later period he portrays the nature of language as primal Saying - the manifesting within the unhiddenness of the clearing of Being, of that which is the manner in which it is.

---

[214]W.J. RICHARDSON, *Phenomenology to Thought*, xxiii.

[215]W. BIEMEL, "Poetry and Language in Heidegger" in J.J. KOCKELMANS, *On Heidegger and Language,* 92-93.

Although Heidegger initially discusses language in BT as an instrument of human expression and admits the value of the studies of grammar and linguistics within the sphere of the sciences, he considers that the scientific treatment of language under the title of linguistics is incapable of grasping the essential nature of language. In his view the essence of language is to be grasped only in a realm which is prior to the realm of the sciences-the realm of the truth of Being: "When ... the truth of Being has become memorable to thought, then reflection on the essence of language must obtain a new rank. It can no longer be mere philosophy of language."[216]

Hence, the later Heidegger arrives at the understanding that the essence of language as primal Saying is the call of Being. Even though the central theme of section 34 of BT is that language as discourse is constitutive of the Being of Dasein, in his later works it becomes evident that the Being of Dasein is the original grant of Being to man. From such a perspective the constitution of Dasein, including language as an intrinsic factor of that constitution is the grant of Being to man whereby man is as that which he is. In accordance with his interpretations of various fragments of the pre-Socratic Greek thinkers Heraclitus and Parmenides, Heidegger maintains that Being has its own fundamental speech, and that this speaking of Being is the fundamental mode of speech. It is evident that later Heidegger considers the nature of language to be distinct from any specific instance of human speech. Nevertheless the essence of language is related to human speech because language needs man in order to bring the message of the call of being into expression. He indicates that all human speech resides within the realm of the essence of language. However most human speech falls short of

---

[216]LH. 222.

the call of Being because it fails to bring that which remains unsaid into expression.

Heidegger's radical critique of metaphysics is at the same time a radical critique of the traditional Western conception of language. Hence his overcoming of metaphysics is at the same time a breakthrough to a new relationship to language. The traditional metaphysical conception of language turns on the distinction between meaning (*Bedeutung*) as something interior and expression (*Ausdruck*) as something exterior. The word is an external sign of a meaning which has been constituted interiorly, in the mind. This essentially Aristotelian conception of language dominated all subsequent theories of language. In this view, language is primarily taken as communication. Since language does not play a role in the formation of meaning, it is assigned the role of communicating meanings which are already constituted. Heidegger rejects this dualist theory and insists that language gives birth to meaning, to the appearance of what appears.

Language is not primarily communication but the way in which things emerge into presence. Nor is language primarily something which man does, rather, it is the doing of Being in man, in human speech. This radically new conception of language is brought out in his essay named *Language* in which he claims that, properly speaking, it is not man who speaks, but language itself. Of course, men do speak, but their speech comes as a response to a more primordial address which issues from Being -or "language"-itself.

Language is not representative, but manifestative. Language manifests, or shows itself in that it makes things light up. Ultimately Being shows itself in language. The showing of the primal Saying of Being is prior to all human speech and is determinative of all human speech. Since primal Saying is prior to and transcends all human speech, primal Saying is wordless and is not subject to expressive utterance. Nevertheless, primal Saying motivates human speech by giving a call to man to come into Being as that thoughtful being which he is. According to

Heidegger human speech is always and essentially a response to the primal call of Being, man's response is most authentically to listen to the call of Being. But as a man listens thoughtfully to the call of Being, he is led to express his understanding of this call to his fellows according to the utterance of human speech. Thus, speech is understood to be derivative of the primal Saying of the call of Being. Hence, language for Heidegger, is not merely an instrument of communication, but is basically the coming of Being into Saying.

GENERAL CONCLUSION

BEING AND LANGUAGE IN BHARTṚHARI AND HEIDEGGER: A SYNTHESIS

1. INTRODUCTION

An attempt has been made to present two distinct philosophies on Being and language evolved from two different cultural contexts. We have analysed Bhartṛhari's understanding of Reality along with his thinking on language as well as Heidegger's question of Being in relation to his understanding of language. We arrive at the conclusion that the question of Reality, Being and language are intimately inter-related in the philosophies of Bhartṛhari and Heidegger. We find the fact that all human problems, being essentially human, reach beyond the bounds of any tradition and confirm the universality of human concerns. Thus, this study of Bhartṛhari and Heidegger points to the universality of thinking.

Our prime concern is not the question of influence of one thought on the other, or one philosopher on the other, rather what is looked for is the independent congruence of the thought content. Attention is focussed on an understanding of the philosophies of Being and language independently of any influence while the universality of thinking is emphasized. No other justification can be offered for such a questionable enterprise of bringing together two different ways of thinking, so wide apart in time in their entire context. Rather than answering any question it could be more a task of trying to see and lay open the hidden truth of the paths taken by thinking from a comparative perspective.[1]

2. *ŚABDATATTVA* - THE REAL AND BEING - THE ABYSS

Much of Indian philosophy exhibits the attitude of philosophizing from the point of view of the human situation. Whatever may be the truth about the conception that Indian philosophy to a great extent is *mokṣa-oriented*, as Mohanty observes, the undeniable fact remains that much of the Sāṃkhya, Vedānta, Buddhist and Yoga systems is not just neutral system-building but is an attempt to think the human situation through. The central metaphysical thesis of these systems springs from the practical interest of leading man out of *duḥkha* into a state of bliss, from inauthentic to authentic existence, from absorption in the object to self-realization. The concepts of *duḥkha* and *Sorge*, as also those of *saṃsāra* and *Geworfenheit*, are closely allied. They designate an ontological situation of passivity, of being subject to causality, of helplessness, etc. They stand for something more fundamental that is constitutive of the

---

[1]As G. Parkes suggests most philosophy practiced with an appropriately historical sensibility involves comparisons. Hence an 'East-West' comparative philosophy is in principle no different from 'comparative' philosophizing within a single tradition. G. PARKES, *Asian Thought*, 2. It is interesting to note that R. May argues in his *Heidegger's Hidden Sources* (London: Routledge 1996) that Heidegger drew upon German translations of Chinese Taoist and Zen Buddhist classics for some of the major ideas of his philosophy. It seems that in his references to the Orient, Heidegger has mainly confined himself to the far East. He has mentioned the Chinese *Tao* as Being, along with the Greek *Logos* and his own *Ereignis*. However, Heidegger hardly ever mentions Indian philosophy. This lack of enthusiasm for the Eastern thought, as J.L. Mehta suggests, may be the result of his pre-supposition that the Eastern thought, being presented in Sanskrit, tends to promote 'metaphysical' thinking which he strived to overcome. J.L. MEHTA "Heidegger and *Vedānta*: Reflections on a Questionable Theme" in G. PARKES, *Asian Thought*, 24.

human predicament.[2] Bhartṛhari shares in the central attitude of Indian thinking that springs from the practical interest of leading man from inauthentic to authentic existence, from absorption in the object to self-realization.

Heidegger also, it seems, expresses his dissatisfaction over the 'thrown fallenness' of every day life. Just as Dasein understands its own being in terms of the beings to which it attends, Bhartṛhari's man through action gets involved and absorbed into the objects of everyday reality. As is the case with Heidegger's preparatory analytic, Bhartṛhari proposes to change the ordinary human condition which calls for a fundamental alteration of attitude that facilitates awareness of how it is with everyday life. The suffering that leads the self to Brahman and the dissatisfaction that leads thinking to *Ereignis* open up paths of such striking similarities that hemispheric differences cannot entirely obscure it. Heidegger in rediscovering for the West man's belongingness to *Ereignis* comes closer to Bhartṛhari's man who dissociates himself from untruth through *Vāgyoga* and achieves union with the Real, *Śabdatattva*. Man, says Heidegger, is the 'Neighbour of Being.' Man of all beings stands in a privileged relation to Being. BT attempts to discover Being in and through a reflection on the nature of Dasein. Similarly for Bhartṛhari an analysis and reflection on the nature of the self is a means of knowing Brahman, the *Śabdatattva*.

Bhartṛhari's *Śabdatattva* is characterized as Being[3], as

---

[2]J.N. MOHANTY, *Essays on Indian Philosophy Traditional and Modern* ed., P. BILIMORIA (Delhi: Oxford University Press 1993), 289-290.

[3]The verb *asti*, having for its root *as* and *sattā*, the abstract derivative of *sat*, are the two terms crucial for any thinking about Being in Sanskrit. The root *as* has, among its meanings, to live, to exist, to be present, to take place, to happen, to abide, dwell, or stay, and also become. Hence the root means, both to be present and to happen as well as to become. *Astitā*, or existence, is then that which is common to them all, existence in a sense which applies both to

distinguished from all beings. It is not one object amongst others; it is that in the light of which everything else is manifested and which is itself self-manifesting. Bhartṛhari seems to consider existence (*sattā*) as a universal substance (*dravya*), a timeless, simple substance underlying all things and permitting them to borrow their existence-claims from their 'association' with it. While discussing the nature of Brahman, Bhartṛhari interprets it as *dravya* (substance). Bhartṛhari considers particular things as real or substantial in the everyday world but they are not real or substantial in the ultimate sense. For him, substance is the universal in particular things which is nothing other than Brahman as existence. Brahman as substance is also called *tattva* (thatness) which is further referred to as *para* (the highest one).[4] As *tattva*, it neither is nor is not; it is neither one nor differentiated; neither combined nor separated; neither changing nor unchanging.[5] This eternal existent is made known to us through nonexistent forms (*ākāra*); likewise, it is this existent alone that is designated by linguistic forms that designate nonexistent limitations (*upādhi*).

The Absolute transcends all spatial and temporal determinations, differentiations based on concepts like identity, difference, being and non-being are not applicable to it. For these reasons diversity and multiplicity could not really be part of the essence of the Absolute. In the same way the various forms of changes can in no way be established in true Reality. The Reality seems to have these forms as its essence but it is something whose

---

that which is present, to a happening or occurrence, and to becoming. The word *sat* is more commonly used in philosophical literature. While it retains the ontologically neutral meaning of *astitā*, it goes beyond that in its connotation so that besides meaning being, existing, occurring, happening present, living, and enduring, it also came to mean the true, the good, and the right.

[4]VP. III.1.20.

[5]VP. III.2.12.

nature is not at all comprised of these forms.[6] This again is an indication to the transcendental and immanent aspects of Reality. The Grammarian emphasized the fact that Brahman as the basis of differentiated aspects is of the nature of *śabda* which he calls *Śabdatattva*, one undifferentiated reality, the whole.

Brahman is *sat* (Being), the ground of all that is, including one's own being, which is of the nature of pure *cit* (consciousness), of which "knowing" is itself a derivative mode and potentially capable of rising above all otherness. Thus Brahman is Being, but not in the sense in which it is other than what it is Being for or to, not in the sense of what knowing and speaking are about, other than them, as a reality confronting them, but inclusive of these as themselves modes of Being. Brahman, being beyond the reach of the senses cannot be an object of consciousness accompanied with the idea of either existence or non-existence. Brahman, therefore is beyond being or non-being. This Brahman whom Bhartṛhari considers as the great universal (*Mahāsāmānya*), the great Being (*Mahāsattā*) expresses itself in all words.

Bhartṛhari conceives the ultimate Reality as One. It is of the nature of word and from it are manifested all objects and the whole cosmos. To explain the emergence of the world of immense diversity out of a single unity, the Grammarian felt obliged to recognize the various powers of the ultimate principle. Bhartṛhari has tried to explain the plurality by having recourse to the concept of *śakti* inherent in the Word-Principle. He treats the *kālaśakti* as the supreme of all the powers of the Absolute because it controls all other powers of *Śabdatattva* by presenting them in a particular temporal order. *Kāla* is not different from Brahman, but a manifestation in the sequence of coming into being. When such time sequences appear as differentiated objects, then time as a

---

[6]VP. III.2.10.

power seems to be different from Brahman, but really it is not. Time is the efficient cause by which Brahman controls the cycles of the universe. Since change takes place only in entities conditioned by space and time, Brahman which is not affected by these categories is also beyond all change.

From the transcendental point of view, Brahman cannot be a real or unreal cause, for it is ineffable. Brahman merely appears as the world of multiplicity under the operation of its powers. Real and unreal are not different entities but real, when not comprehended properly, is believed to be unreal.[7] Time as an inherent power of the Absolute exerts its influence in bringing about the powers of *Śabdatattva*.[8] It does not mean that the *Śabdatattva* is limited by Time, but the eternal timeless appears as changing owing to the working of time factor.[9] This undifferentiated and unchanging Reality, beyond determinations of thought and words is non-relational as there is no other entity with which it could be related or associated. It is immutable as it transcends space, time and causality. It has neither a beginning nor an end.

Bhartṛhari also refers to that aspect of the eternal reality which transforms itself into speech, as words and their meanings and into matter as the different subjective and objective elements of the phenomenal world. Nevertheless, these transformations do not affect the true essence of Reality and it remains unchanged while persisting in all its modifications. Bhartṛhari declares that there does not arise any cognition which is not fused with *śabda*. All knowledge appears intertwined with *śabda*.[10] Hence the object is really not different from the word. It is the word which has

---

[7]VP. III.2.7.

[8]VP. I.3.

[9] K. RAJA, *Meaning Theories*, 146-147.

[10]VP. I.123.

become the object.[11] The world of the knower exists only as presented by *śabda* or as the meaning of *śabda*. So for Bhartṛhari all activity of consciousness has *śabda* as its medium. If knowledge were to cease to be of the nature of *śabda*, consciousness would remain unmanifested to one's own self.[12]

Bhartṛhari proposes *Vāgyoga*, meditation on the unity of the ultimate Reality which is Word, to reach the complete absence of all sorts of differentiation and sequence. *Vāgyoga*, therefore, is a kind of meditation which aims at raising the level of the consciousness of words to the highest stage of the Word-Principle. The *Vāgyoga* demands a kind of *śabdasamskāra,*[13] i.e., the purification of words. This purification consists in discarding of unchaste or corrupt words (*apabhraṃśa*) and adherence to chaste or correct words (*sādhu śabda*) which culminate in the attainment of the ultimate Reality.

Now Heidegger, as the philosopher of Being explains his own philosophy as a rethinking of the traditional metaphysics of being. He proposes meditative thinking as a new way for his *Seinsdenken*. Being for Heidegger is always Being as it enters into Dasein's understanding of Being. Being the *Da* (there) of *Sein* (Being) man is the Da-sein.[14] Man's ek-sistence is ek-static openness to Being. This ek-static standing of Dasein is also a standing out in the truth of Being.[15] The *Da* in Dasein is Being itself revealed or brought out of concealment into disclosure. Since disclosure in Heideggerian thinking is truth, Dasein is the truth of Being. Dasein is the place of the truth of Being. Dasein being the "There" (presence) of Being; human existence is, in the final analysis, the openness to Being, "the standing in the lighting-

[11]K.A.S. IYER, *Bhartṛhari*, 100.

[12]VP. I.124.

[13]VP. I.132.

[14]BT. 171.

[15]LH. 229.

process of Being" (*das Stehen in der Lichtung des Seins*).[16] The human being dwells in the light of Being; the lighting-process (Being) "grants the nearness to Being."[17]

The relationship between Being and man cannot be thought of in terms of the subject-object relation. Since Being is in no way a being, but is rather not-a-being, it is Nothing. This Nothingness of Being makes Dasein conscious of its inauthentic state of existence and invites it to authentic existence. To exist authentically, Dasein should understand the true nature of Being; and to understand the true nature of Being, Dasein has to understand itself, because it is only in Dasein that the true nature of Being is revealed. An untrue understanding of Being may leave Dasein in an inauthentic state of existence. Being in unconcealment as truth presents itself in Time.

As we have seen, something like primordial Time shows a predominant concern for Being's manner of occurrence. To be sure, primordial Time remained "tied up" with man; even though 'Time is not the product of man and man is not the product of Time.'[18] After having learned that time cannot be regarded most basically as the structure of human understanding Heidegger tries to think Being as somehow determined in its presencing by Time. As to the relation between Time and Being, Heidegger fixes on the notion of presence. Being and Time says Heidegger, determine each other reciprocally, but in such a manner that neither can the former, Being, be addressed as something temporal, nor can the latter, Time, be addressed as a being.[19] He continues: "From the dawn of Western-European thinking until today, Being means the same as presencing (*Anwesen*)." Thus, somehow, he puzzles, "Being is determined, as presence, by

---

[16]LH. 229.

[17]LH. 223.

[18]TB. 17.

[19]TB. 3.

Time."[20] The present task, says Heidegger, is to think Being and Time together.

The Real and Time belong together. How can this be? How can Real permit both the whole world and Time to be what they are, and yet only by not becoming them? Heidegger suggests that "in Being as presence there is manifest an ongoing arrival which so concerns us as humans that in growing aware and accepting this arrival, we attain the distinguishing mark of human being."[21] To accept this arrival is to stand within the open field of giving so that the interplaying dimensionality of genuine Time reaches us. And because Being and Time are only there and given in *Ereignis*, it is this event that brings man into his own as that being which grows aware of being by standing in genuine Time.

In the expression *es gibt Sein, es gibt Zeit,* Heidegger finds a double way of thinking this reciprocal determination: on a primary consideration, "there is Being, there is Time," this expression turns thought from metaphysical representations of their relationship -by reminding us not to think Being and Time as things or the ground of things, or as thing static that happens or dialectically unfolds. But then, taken to say "It gives Being, it gives Time," this expression turns thought toward a cautious meditation upon the "it" which gives by letting Being be brought into the unconcealedness of presence. In the process of unveiling, there is a kind of granting at work which grants Being-present, while it lets-be-present that which is present, namely, beings. In this process we come again upon a granting, and thus upon an "It" which grants. Heidegger seems to name the 'It gives' as *Ereignis*. In the event man comes to himself, discovers his identity and relationship with the Being. *Ereignis* is therefore the name given by way of anticipation to a possible 'identification' between

---

[20]TB. 2.

[21]TB. 23.

man and Being, beyond metaphysics.

From an ontological perspective, Brahman stands for true Reality (Being). Brahman as true Being makes the truth of the world possible. A reflection on the Reality of Brahman in philosophical terms, brings us closer to Heidegger's concept of Being. Because Being is no-being (no-thing), it cannot be described as such or such. It is non-being as Radhakrishnan points out, since it is not the being which we attribute to the world of experience.[22] Heidegger considers Being as the groundless ground and Abyss of all human understanding, but the full Truth of Being is not accessible even to the highest form of insight. Because the Truth of Being is forever hidden from us, Being in the last analysis remains a secret.

Heidegger's analysis of There-being was to serve the purpose of bringing to light Being which illumines beings. The 'There' of There-being is only in the clearing. There-being transcends toward Being. Moreover, There-being and Being stand in a reciprocal relation. Being conceals and reveals itself in the world as what Heidegger calls epochal e-vents. When Heidegger calls man the 'shepherd of Being' (*Der Mensch ist der Hirt des Seins*), he wants to indicate the special relation There-being has to Being. From Bhartṛhari's point of view also man alone is endowed with the capacity to envision Brahman and thus raises beyond all living creatures.

Bhartṛhari denies a causal nexus between Brahman and the world on the ground that the effect, the world, is a mere appearance (*vivarta)*of the cause. Heidegger also rejects the thought of Being as the cause of the world. For Bhartṛhari and Heidegger cause and effect, if one uses such concepts at all in this context, are intertwined, albeit not in the same manner. Being and Time according to Heidegger belong together. They determine

---

[22]S. RADHAKRISHNAN, *Indian Philosophy*, 2 vols. (Delhi: The Oxford University Press 1997), II. 538f.

each other reciprocally, but in such a manner that neither can Being be addressed as something temporal nor can Time be addressed as Being. Time, *kāla* for Bhartṛhari, is the power of *Śabdatattva*. They belong to each other. *Kāla* is not different from Brahman but is that aspect of Brahman which follows manifested sequence to come into being. Heidegger, it will be remembered, often speaks of the Light of Being, or clearing which brings Light. According to Bhartṛhari Being shines forth in the breaking forth of the word. In Heidegger's account of the belongingness of man and Being, in the Self-same, one hears an echo of the notion of *advaita* which Bhartṛhari upholds.

The crucial and culminating point in Heidegger's thinking is how Truth (*Wahrheit*) itself can be understood in terms of the profounder notion of *alētheia* or unhiddenness. In the Upaniṣadic tradition where from Bhartṛhari draws inspiration, Truth (*satya*) and Being (*sat*) are even more intimately connected than in the Greek. *Sein* in its innermost nature is *Sage*. Being is *Logos*. *Sat* is *satya*. *Satya* as the claim of Being is the *Sphoṭa* emerging from *Sat*. When *Sphoṭa* is listened to in silence, it is *Śabda* - the *adiśabda* (primordial utterance) which is at the same time *anāhata śabda* (unmanifest utterance). In the eventing of Being happens Truth and Dasein becomes the place of the truth of Being. Being reveals, comes-to-pass in being and being dwells in the light of Being. The Real, *Śabdatattva* expresses itself through the breaking forth of words. It shines forth as the *satya*, truth in everything.

3. *SPHOṬA* OF *ŚABDATATTVA* - *SAGĖ* OF BEING

Words that we speak, says Bhartṛhari, come out as an embodiment of thought, which then, is called language. The essence of Bhartṛhari's view lies in the conception that the Absolute Existence itself manifests in the form of words and their meaning. There is no difference between them; Brahman Himself is the word. Our speech imperfectly imitates this Absolute word.

Language, according to the Grammarian, not only reveals reality, but it is reality. Whatever is, is capable of being named. The unnameable is the non-existent. Thus there is an intimate relation posited between thought and language. All the things that comprise reality are of the nature of the word because they are apprehended as identical with it in all one's cognitions. The whole phenomenon of material existence is only an appearance of the *Śabdatattva* which is identical with the ultimate reality, Brahman. Thus language itself is the Reality.

Real appears as words and as objects denoted by them, though this distinction is but a convenient fiction. According to Bhartṛhari speech and thought are but two aspects of the same Speech-Principle.[23] The sentence is the fundamental linguistic fact and letters and words are unreal abstractions from it. One primary problem faced by Bhartṛhari was how the passing sounds constituting a word could have a single meaning for the listener. He postulated a meaning-bearing symbol called *sphoṭa*, behind the several phonemes forming a word, or many words making up a sentence. It was the eternal meaning that was communicated by way of suggestion through sounds. *Sphoṭa* in its literal sense is normally defined as 'that from which the meaning bursts forth,' i.e., shines forth, in other words the word as expressing a meaning. The basic logic behind his thinking is that the whole is prior to the parts.

What Bhartṛhari is emphasizing is that the *sphoṭa* is not merely restricted to the transitory sounds.[24] There is also a certain order in which the suggester and the suggested are realized. For him, the eternal *sphoṭa* becomes concretized in the suggestiveness of the sounds and the words.[25] An important factor in this meaning-

---

[23]VP. II.31.

[24]VP. I.95.

[25]VP. I.91.

bearing function of the sounds and words is that it is along with the suggested that the suggester is revealed. It is through the perceived sounds that the meaning behind it is grasped. Since the utterance of a statement took several seconds in which the earlier phonemes would be gone and lost before the latter ones were uttered and heard, there should be a single symbol, according to the Grammarian, to account for the continuity and unity of the whole utterance which he named *sphoṭa*.

Bhartṛhari argued that words, concepts and universals are all constructs. Words are abstracted out of whole sentences or sentence-complexes, and similarly concepts are abstracted out of whole. This abstraction is done through our inherent faculty of speech-impregnated awareness. The reality is an impartite whole, which we never cognize as such, but always under the guise of some words or word-generated concepts. This means that particulars are abstractions or are falsely broken or extracted pieces of the indivisible whole. For Bhartṛhari, having a concept and using a word are merely the two sides of a coin. They arise from the same capacity of the mind. Bhartṛhari therefore asserts: 'There is no awareness in this world without its being intertwined with the word. All cognitive comprehension appears to be penetrated as it were, with the word. If the property of being impregnated with speech (word), which eternally belongs to comprehension, were purged out of it, then (its) illumination would not illuminate. For this (word-impregnated nature) is the intrinsic activity of consciousness.'[26] These verses point to the essential and fundamental linkage between language and thought.

As Matilal points out, for Bhartṛhari even our awareness of particulars that pertain to the so-called immediate experience is 'intershot' with language.[27] Even in the pre-linguistic cognitive

---

[26]VP. I.123-124.

[27]B.K. MATILAL, *Perception* (Oxford: Clarendon Press 1991), 388.

state in Bhartṛhari's view there lies a speech-latency, i.e., a completely inert speech in latent form. In other words, it is cognitive, then it has speech-potential. The speech-potency is an essential trait of human consciousness. Even babies would have it. "This (speech-potential) is certainly the consciousness of all beings; it exists within and without. The awareness of (sentient) beings does not transgress this (speech-potency)."[28]

Martin Heidegger has developed his understanding of the nature of language in relation to his philosophical problematic of the question of Being. "Language is the clearing-and-concealing advent of Being itself."[29] Language becomes the concrete presence of Being rather than mere articulation. In as much as language grants to man this capacity of speech Heidegger claims that language is "the foundation of human being."[30] The essential being of language is the *Sage* as that which shows [31] (*Des Wesende der Sprache ist die Sage als die Zeige*). The showing of Saying takes place when language breaks in to speech. It is language itself which reaches into the regions of presencing and lets what-comes-to-presence appear and disappear.

The primal Saying of language which is prior to all human speech is a monologue.[32] This monologue is characterized as "stillness (*Stillen*)."[33] Language needs and uses the speaking of humans in order to bring the primal Saying to expression. Man belonging to Saying (*der Sage gehörend*) listens to it and its word and brings what it hears "correspondingly" into human-sounding words. Mortal speech is thus characterized as a co-respondence

---

[28]VP. I.126.

[29]LH. 230.

[30]LH. 230.

[31]WL. 123.

[32]WL. 134.

[33]PLT. 207.

(*ent-sprechen*) or a response (*Nachsagen*).[34] "Thought is attentive to the clearing of Being whenever it inserts its Saying into language. ... In this way thought is a doing (*Tun*)."[35] Thinking acts to the extent that it "brings Being to language." (*Zur Sprache bringen*). Language as Saying is the mode of appropriation (*Ereignis*). [36]

Heidegger has clearly affirmed the fundamental role of language in constituting the world and our experience. Language is not "the means to portray what already lies before one", but rather it " grants presence - that is Being -wherein something appears as existent."[37] Since language is "the house of Being," one reaches Being by constantly going through this house. " Whenever we go to the well, or walk through the wood, we are always already going through the word 'fountain' and the word 'wood,' even though we are not saying these words or thinking of anything linguistic."[38] In saying this, Heidegger means that language is correlative with experience of the world. Phenomena in the world occur simultaneously with the occurrence of language and the world exists only where words exist. Language is that mode of the Being of man which permits him to be open to the beings that relate themselves to him. 'Language, writes Heidegger is not only a tool which man possesses among many others, but language first grants the possibility, in general, to stand in the midst of the openness (*Offenheit*) of beings.'[39]

In Heidegger's view, thinking and language are inseparable. We "always already" find ourselves in the midst of language: we are attuned to language as reality rather than as a surrogate of

---

[34]WL. 134.

[35]WM. 239.

[36]WL. 135.

[37]WL. 146.

[38]PLT. 132.

[39]HEP. 276.

reality. To think of language is to think of reality. Jung calls man's attunement to language "linguisticality" (*Sprachlichkeit*) -that which makes thinking possible.[40] It is by means of the possession of language that man is not closed off from that which reveals itself but can come under influence and illumination. The fact that verbal constructions are in a position to communicate states of being is not an indication that communication is the basic function of language. The communicative function of language is a result of the fact that Being reveals itself in language. Language is therefore the ground in which the Being of man is founded.

Heidegger quotes Hölderlin to the effect that man is a conversation.[41]Here Heidegger makes a direct connection between language and Being according to which man is no longer to be thought of as disposing of language as he wills, as if it belonged to him as his own most cherished means of expression: "Thinking accomplishes the relation of Being to the essence of man. It does not make or cause the relation. Thinking brings this relation to Being solely as something handed over to it from Being. Such offering consists in the fact that in thinking Being comes to language. Language is the house of Being. In its home man dwells. Those who think and those who create with words are the guardians of this home. Their guardianship accomplishes the manifestation of Being insofar as they bring the manifestation to language and maintain it in language through their speech.[42] Being has the primacy over thought, simply because it gives rise to it.[43] Heidegger says: "Said plainly, thinking is the thinking of

---

[40]H.Y. JUNG, "Heidegger's Way with Sinitic Thinking" in G. PARKES, *Asian Thought*, 222.

[41]HEP. 277.

[42]LH. 217.

[43]H.G. Gadamer, one of Heidegger's many eminent students, discusses the primacy of language and develops in his *Truth and Method* (London: Sheed & Ward 1975) the linguisticality of human experience. Along with

Being. ... Thinking is of Being in as much as thinking, coming to pass from Being. At the same time thinking is of Being insofar as thinking belonging to Being, listens to Being."[44] It implies that thought belongs to Being (as Being's attend-ant), for it is brought about by Being; that thought attends to Being because it comes about in There-being.

There is a deep kinship in the philosophical outlook concerning the question of language in Bhartṛhari and Heidegger. For a long time he has searched, Heidegger tells his Japanese partner in a conversation in "Dialogue"[45] for a word to distinguish primordial from other kinds of speech. He has finally found it in *die Sage* (primary utterance), derived from the ancient German word *sagan*, meaning to show, to let be and see. *Die Sage* precedes all ordinary languages because before man utters words he receives a message of or is 'shown' what 'is.' Speech captures precisely that which has come into being and speaks in the silent mode of primary utterance.

For Heidegger the primal Saying prior to all human speech is a monologue which might correspond to Bhartṛhari's *sphoṭa* in the unuttered stage of speech as the indivisible whole. Now the primal Saying is brought to expression through human speaking in the same manner *sphoṭa* breaks in to words through human speech. It is thinking that brings Being into language and Bhartṛhari finds the speech potency as an essential trait of human

---

Heidegger, Gadamer's contention is that 'coming to language' (*Zur-Sprache-Kommen*) constitutes a universal ontological structure and, therefore, all that can be understood is language.

[44]LH. 220.

[45]The "Dialogue on Language between a Japanese and an Inquirer" (1954) is the only extended engagement of Heidegger with Asian philosophy. Heidegger has pondered the being of language as deeply as any other thinker, and yet he has his questioner admit near the beginning: "I do not yet see whether what I am trying to think as the essence of language is also adequate to the nature of East Asian language ..." WL. 8.

consciousness. In Heidegger's view phenomena in the world occur simultaneously with the occurrence of language. The world exists only where words exist. This seems to be what Bhartṛhari meant when he says 'all things that comprise reality are of the nature of word.' For whatever is, is capable of being named.

Besides the striking similarities, the two philosophies diverge radically. For Heidegger, even *Sein,* and not merely *Seiendes* is temporal, historical, and in that sense finite. Heidegger frequently speaks of the destiny of Being, but all such expressions become simply meaningless within the system of Bhartṛhari. For Heidegger, Being is 'the process by which finite beings emerge from concealment.' For Bhartṛhari, the process of emergence from concealment is a process of destruction of that which apparently conceals. The Being which itself is self-manifesting, manifests all others including concealment itself. In the eventing of Being, it seems, Heidegger is not distant from Bhartṛhari moving from finite to the unspoken realms of Being.

Heidegger, in an attempt to think about the Western thought, characterizes it as metaphysical beginning with Plato and culminating in Nietzsche -a tradition which has led, in Heidegger's view to technology as a way of will to power, of objectifying and calculative thinking. Heidegger believed that with the end of modernity, philosophy has come to an end - heralding the need for a new kind of thinking that would not be objectifying or metaphysical. Contrary to Heidegger's characterization of Western thought as destined to be a metaphysics of subjectivity, Indian thought in general and Bhartṛhari in particular speak in terms of the identity of subject-object realms reaching beyond all sorts of distinctions and differences.

In the process of thinking the Being Heidegger seems to move towards mysticism. Yet he has made his position clear that as a philosopher he is not concerned with religious problems, while Bhartṛhari does not seem to make a distinction between religious and strictly philosophical speculations. For him, knowing

Brahman (though he does not identify it as God) through *Vāgyoga* can lead one to attainment of *mokṣa*. God in Heidegger's view would be quite different from Brahman. Heidegger realizes that the way to deal with the transcendent and "simply other" reality (of God or Being) is not to deal with it at all, but to let it deal with us. The only way to gain access to Being itself is to let Being be and to let it address man.

4. CONCLUSION

The question of Being is a bridge that links the traditional philosophy to the philosophy up to our times. While every other determination can be applied to some things and not to others, Being, in one form or another, can be predicated of everything that is. This is a result of the all-pervasive character of Being. All things are and can be Being and a form of Being. To the extent, therefore, that things are, they are unified by that which is common to all of them, that they are Being. It serves as a unifying aspect of everything. In this sense, Being constitutes the unity of everything that is. This unity forbids change and diversity because these are incompatible with the total unity that Being implies.

Heidegger makes it clear at the outset of his thinking that if concern with the problem of Being implies, as its aim, a drawing up of a definition of Being, then disappointment is inevitable. The reason for this is that 'Being, for purposes of definition, cannot be deduced from higher concepts and cannot be presented by lower.'[46] The question of Being, as he raises it, amounts at the same time to a quest for a way of thought that can redeem our humanity by overcoming "the oblivion of Being" and so enable us to break the omnipotence of technology and live in the world as truly our home, in the "neighbourhood" of Being. Heidegger has made it amply evident in what sense Being is ground-word of the Western tradition, and how Being is the guise in which reality

---

[46]BT. 23.

disclosed itself at the beginning of this tradition and why it remains a question supremely worthy even today.

In similar fashion, one may say that Brahman and *Ātman* are the ground-words of the Indian tradition which Bhartṛhari shares as the very embodiment of the primordial unhiddenness of *Śabdatattva*. An inquiry into the truth of this Being without the metaphysical mask challenges us with the *parāvidyā*, the higher knowledge, for a deeper comprehension of our own being. Ek-sisting to Being - *Śabdatattva*, being receives the light of Being, unveiling the truth of Being. Then the being speaks the primordial *Sage* - *sphoṭa* which breaks forth as Being. In this deeper unity of the Being - *Śabdatattva*, being finds its authenticity.

In an age of constantly narrowing distances between nations it is most important that East and West come to a deeper appreciation of their respective intellectual commitments and communicate it in search of Being beginning with being. Heidegger's path of thinking the Being and Bhartṛhari's meditative way of knowing the *Śabdatattva* lead one to the belonging togetherness of Being and being in a deeper unity which is a continued process beyond all conceptualizing, transcending the bounds of Orient and Occident.

GLOSSARY OF SANSKRIT TERMS

Ābhāsa:	appearance
Abhihitānvaya:	relating the meanings of words of a sentence after they are uttered; opposed to *anvitābbidhāna*
Adiśabda:	primordial utterance
Advaita:	non-dualism; monism; system of Vedānta associated with Śankara
Advaitin:	non-dualist; monist
Āgama:	that which has come down by tradition; Revealed Scripture; any sacred text
Aham:	ego; the notion of 'I'
Ajñāna:	ignorance; absence of knowledge
Ākāra:	form
Akṣara:	imperishable, non-transient; the imperishable, the non-transient; Brahman as God
Anāhataśabda:	unmanifest sound
Ānanda:	bliss; blissful; an essential attribute of Brahman
Ananta:	infinite; that which is not conditioned by space, time or another entity; a distinguishing attribute of Brahman
Anubhava:	experience
Anumāna:	inference
Apabhraṃśa:	corrupt words
Apauruṣeya:	not ascribed to human author
Apavarga:	liberation, salvation
Artha:	meaning; significance; object; wealth; purpose, aim

Asat: non-being, non-existence; the non-existent, the unreal; false, untrue

Asatkārya-vāda: the doctrine that the effect is non-being before it is born

Ātman: the self; the individual soul

Avidyā: ignorance; the cosmic principle which is the cause of the world illusion

Brahmā: the Vedic deity entrusted with the task of creation of the universe

Brahman: the Absolute of philosophy; the ultimate Reality; the personal God according to many forms of Śaivism and Vaiṣṇavism

Brahmānubhava: experience of Brahman

Cit: consciousness

Darśana: vision; a system of Philosophy

Dharma: that which supports; nature; the law of nature; virtue, ethical law; the "ought"; merit; the potency of ethical actions; the right action, the law or the body of the doctrines of any faith

Dhātu: original element

Dhvani: the function of suggestion, suggestive expression or meaning, suggested content; poetry with principal suggestion

Dik: space; direction

Dravya: a substance; that which is associated with modification

Duḥkha: sorrow, suffering

Dvaita: dualistic; a system of Vedānta founded by Madhva

Eka: one; one Reality

Gauṇa: secondary

Guṇa: quality; property; attribute

Guru: preceptor imparting spiritual knowledge

Lakṣaṇā:	secondary sense
Jāti:	genus; universal; source
Jijñāsa:	enquiry into the nature of Brahman, desire to know Brahman
Jñāna:	knowledge; consciousness
Kāṇḍa:	section
Kāla:	time
Kālaśakti:	time power
Kāraṇa:	cause
Kārikā:	verse
Karma-kāṇda:	the earlier part of the *Vedas* dealing primarily with the rituals
Karma:	action, rituals; past deeds and their results in the form of merit and sin
Kārya:	effect
Kavya:	poem, poetry
Khaṇḍapakṣa:	analytical method
Maṇḍala:	a division of the *Ṛgveda* (the whole collection being divided into ten maṇḍalas)
Mantra:	esoteric syllables or words signifying spiritual ideas, a Vedic hymn
Mahāsāmānya:	the great universal
Mahāsattā:	the great Being
Māyā:	cosmic principle which gives rise to world illusion; the phenomenal character of the universe; the primordial matter (*prakṛti*)
Māyā-vāda:	the doctrine of Advaita Vedānta which advocates that everything other than Brahman is illusory
Mīmāṃsa:	discussion; debate; criticism; critical interpretation; name of the school of Jaimini, based on the first two parts of the *Veda*
Mokṣa:	Liberation; salvation; emancipation; a

	complete and comprehensive experience of Brahman; attainment of Brahman
Mukti:	same as *mokṣa*
Nāda:	sound
Nādabrahman:	the sound Brahman
Nidāna:	cause
Nimitta-kāraṇa:	instrumental cause
Nirguṇa:	without qualities, devoid of all characteristics, without attributes
Nirukta:	a treatise containing glossarial explanation of vedic terms
Nitya:	eternal
Pada:	word
Padārtha:	word import
Para-brahma:	the ultimate metaphysical Reality; the supreme personal God
Para:	the highest; the Supreme
Paramartha:	ultimate, highest, absolute; the absolute reality or truth
Paramātman:	the Supreme Spirit
Pariṇāma:	change, transformation; transformation in which the being of the material cause is changed into something else and cannot be regained; modification; evolution
Prāṇa:	vital breath; also applicable to Brahman
Prakāṣa:	illumination
Prākṛtadhvani:	primary sounds
Prakṛti:	primordial cosmic matter
Prāmāṇas:	means of valid knowledge
Pratibhā:	vision; genius; intuition; creative imagination
Pratyakṣa:	perception
Śabda:	sound; word; in grammarian philosophy, the same as *sphoṭa*; verbal testimony

Śabdabrahman:	word-Absolute
Śabdapṛamāṇa:	verbal testimony
Śabdasaṃskāra:	the purification of words
Śabdatattva:	word -Absolute
Śabdavyāpāra:	the power of words to communicate meaning
Sādhana:	method or means adopted to achieve a goal
Sādhu śabda:	correct words
Saguṇa-brahman:	Brahman endowed with attributes
Śakti:	energy, power, potency; the energy aspect of the Godhead. The supreme female deity; import of a word
Sambandha:	relation
Saṃsāra:	the flux of the world; the flow of the world; world
Saṃskāra:	latent impression; instinct; imprint; innate tendency; innate potency
Samuddeśa:	chapter
Śāstra:	sacred texts
Sat:	being, existence, reality
Satkāraṇavāda:	the doctrine that only the material cause, but not the effect is real; the doctrine that Being is real, its forms and shapes are not real
Satkārya-vāda:	the doctrine that the effect is real or existent even before it is born; the doctrine that the effect exists in a latent form before it is born
Sattā:	Being, Existence; derivation from sat; power, force
Satya:	truth; reality; an essential attribute of Brahman
Sāyujya:	the state of *mokṣa* in which the individual

	self enjoys equal status with Brahman
Śiṣya:	disciple
Smṛti:	what is remembered; remembered sacred text like the Epics, *Purāṇas*, *Dharma-śastras*, *Āgamas*
Sphoṭa:	the original seed state of the word before its letters are uttered in a sequence; at that time it is of the form of cognition comprehending the object also
Śruti:	what is heard; a synonym for the *Veda*; Revealed Scripture
Sūtra:	an aphorism; a short, concise technical sentence
Tattva-jñāna:	knowledge of Reality
Tattva:	ultimate Reality; metaphysical category
Upādāna-kāraṇa:	material cause
Upādhi:	adjunct; a limiting condition
Uttara-Mīmaṃsa:	the Vedānta system dealing with Brahman
Uttarakāṇḍa:	the later part of the *Vedas* dealing with Brahman knowledge; the *Upaniṣads*
Vācya:	the expressed or denoted sense, conventional meaning
Vāda:	discussion of a philosophic topic or issue; dialectical argument
Vāgyoga:	meditation on the unity of ultimate Reality which is the Word
Vaikṛtadhvani:	secondary sounds
Vāk:	speech; synonymous with Lakṣmi as *Vāg-devī*
Vākya:	sentence; scriptural statement
Vākyārtha:	purport of a sentence
Vidyā:	knowledge, learning; education; consciousness
Vijñāna:	knowledge: philosophic wisdom

Vikalpa:	conceptualization; empty verbal cognition without a corresponding object; idea
Viśiṣṭādvaita:	non-dualism in the sense of organic unity; a system of Vedānta expounded by Ramanuja
Vivarta:	transformation; transformation in which the nature of the material cause is not affected; distinguished from *pariṇama*
Vṛtti:	commentary
Vyākaraṇa:	grammar
Vyakti:	particular
Vyañjakatva:	suggestiveness or suggestivity

APPENDIX

AN OVERVIEW OF THE NOTION OF *ŚABDA*

In India speculations on the notion of *śabda*, speech or language originated in the vedic times. The Indian seers were interested in the problems of language just as they were pre-occupied with their search for the meaning of the various realities of the universe. Their discussions included many subtle problems related to the origin of a word, different forms of speech, relation of a word to its meaning, speech and its created or uncreated nature, speech and its divided and undivided state etc. The vedic *Saṃhitās*, *Brhmaṇas*, and the *Upaniṣads* as well as *Niruktas* (etymology), the *Praātisākhyas* (Phonology) and the extensive grammatical literature all offer us valuable information about all these problems. This fact, no doubt, points to the keen interest and assiduity of Indian teachers concerning their interest for the study of the phenomena of *śabda* or *vāk*.

We have great works like Yaska's Nirukata, Pāṇini's *Aṣṭādhyāyī*, Patañjali's *Mahābhāṣya*, Jaimini's *Mīmāṃsā* and its commentaries of Śabara, Prabhākara, Kumārila, Bhartṛhari's *Vākyapadīya* and such other master pieces about the investigation into the meaning of language. Though, the Indian speculations on *śabda* were mixed up with religious and spiritual conceptions, they were in no way less rigorous in their scientific and metaphysical discourse on the problem of meaning of the word. This may be the reason why Frauwallner considers achievements of Indians in the sphere of linguistic speculations as their greatest scientific performance.[1]

The Indian Philosophical systems also made many important observations on linguistic problems. Each of the schools of thought had to consider *śabda* at some stage or other, and they had to ponder over some fundamental questions concerning the relation of the word to reality and the validity of verbal knowledge. Except the Cārvākās, the Vaiśeṣikas, the Buddhists, all other systems of Indian Philosophy accept *śabda* as a distinct source of knowledge (*śabdapṛamāṇa*).

The term *śabda* is a very meaningful word. It has a variety of meaning. Therefore it is difficult to be translated into another language by a single term. The varies English terms used to render *śabda* often fail to convey its real sense. *śabda* is generally rendered by terms like word, speech, sound, noise etc. It also connotes a phoneme, vowel, resonance, language and meaning. Stutley's Dictionary of Hinduism gives a fourfold classification of *śabda* namely, 1. *sphoṭa*- a self existent and eternal element beyond perception like meaning; 2. *nāda*- a sound perceptible only to the divinely inspired poet or seer; 3. *anāhata*- the unmanifest and potentially existent element like thought; 4. *āhata*- sound of any kind, whether or not within the range of human or animal perception.[2] However, *śabda* in its literal sense stands for a word (*pada*); and etymologically it signifies sound (*dhvani*). Epistemologically *śabda* refers to the source of knowledge, namely, verbal testimony (*śabdapṛamāṇa*).

An attempt is made here to present a brief exposition of the

---

[1]E. FRAUWALLNER, *History of Indian Philosophy* (New York: Harper and Row Publications 1977), 256. "This grammar," wrote Leonard Bloomfield about Paṇini's grammar of Sanskrit, "is one of the greatest monuments of human intelligence." L. BLOOMFIELD, *Language* (Delhi: Motilal Banarsidass 1994),11.

[2]MARGERAT & J. STUTLEY, *Dictionary of Hinduism* (New York: Harper and Row 1977), 256.

notion of *śabda* as has been traced in the *Vedas*, *Upaniṣads*, the Mīmāṃsā schools and in the Grammar school. Certainly it is not an all inclusive study on the notion of *śabda*. Nevertheless it will provide us with the basic conception of *śabda* in the Vedic tradition.

1. *VĀK*: THE *ŚABDA* IN THE *VEDAS*

The term *vāk* seems to take the place of *śabda* in the *Vedas*. *Vāk* (*vāc*) means word, the sacred, primeval and creative word. It also means sound or speech and language. There are numerous hymns in the *Vedas* which directly or indirectly refer to the notion of *vāk* (speech). Often the whole corpus of the *Vedas* is designated by the name *vāk*, the word meaning revelation. It is also referred to as "eternal word" (*nityā vāk*)[3] in some hymns. Another hymn takes *vāk* as feminine divine principle (*devī*) of the supreme power.[4] *Vāk* being of eternal character, it is not possible to trace back the origin of language. So the people of Vedic times began to think of speech as a creation of God and as the highest gift from God.[5] Hence it remains as an enigma to ascertain when and how man first began to speak. Those who understand the language know the ins and the outs of speech, because the speech unfolds her secrets to them like a beautiful maiden garbed in colourful dress.[6]

The *Ṛgveda* further speaks of a fourfold division of speech. These divisions are known to learned Brahmins; ordinary people do not know the three stages of speech because they are hidden

---

[3] *Ṛgveda*, VIII. 75.6. All the references from *Ṛg Veda* are taken from R.T.H. GRIFFITH, The Hymns of *Ṛgveda*, Vols. I & II (Varansi: Chowkhamba Sanskrit Series 1963).

[4] *Ṛgveda*, X. 125.

[5] *Ṛgveda*, X. 125.3.

[6] *Ṛgveda*, X. 71.4.

from their eyes, they speak only the fourth speech.[7] Yet, another hymn speaks of *vāk* as permeating the whole creation.[8] The Vedic tradition goes further by identifying *vāk* with the Divine. K. Luke confirms this fact when says:

> *Ṛg Veda* embodies a rather subtle conception of the complex reality that we designate as 'word,' and the crucial point in our investigation has been the vedic sages' identification of the word- in the concrete, in their poetical creations- with brahman. We can therefore say that in the perspectives of the poets of the vedic age *vāk*- is brahman.[9]

The *Ṛgveda* states that there are as many words as there are manifestations of Brahman.[10] The *Araṇyakās*, *Brahmaṇās* and the early *Upaniṣads* continue to equate speech and Brahman. Various symbols are used to indicate the divine nature of speech and its evolutionary process leading to creation. Prof. T.R.V.Murti, observes: "The Brhmaṇical tradition stemming from the *Vedas* takes language as of Divine origin (*Daivī Vāk*) as spirit descending and embodying itself in phenomena, assuming various guises and disclosing its real nature to the sensitive soul."[11] Raimundo Panikkar adds to this observation when he states :

> *Vāc*, the word is not just a man made invention of a mere tool of communication or even simply an expression of what man is. The vedic Word is indeed this but is infinitely more. It is ultimately as important as Brahman and in a way that has to be properly understood, it is Brahman itself...[12]

---

[7]*Ṛgveda*, X. I.164.45.

[8]*Ṛgveda*, X. 114.8.

[9]K. LUKE, "Some Aspects of the Rigvedic Conception of '*Vāk*'," *Jeevadhara* 1/2 (1971) 188. Chettimattam observes that "original meaning of 'brahman' was speech coming from the same root as 'logos'."J.B. CHETTIMATTAM, *Experience and Philosophy* (Bangalore: Dharmaram Publications 1996), 169.

[10]*Ṛgveda*, V. 10.2.

[11]T.R.V. MURTI, "Some Comments on the Philosophy of Language in the Indian Context," 324.

[12]R. PANIKAR, *Vedic Experience* (London: Darton, Longman & Todd 1979), 89. Hereafter R. PANIKAR, *Vedic Experience*.

Panikkar presents *vāk* as the total living Word. For *vāk* is more than merely meaning or sound devoid of sense, she is also more than a vehicle of certain truths. She does not merely contain revelation, she is revelation. She was at the beginning. She is the whole of *Śruti*. The *Śruti* is *vāk*.[13]

2. *AUM*: *ŚABDA* OF THE *UPANIṢADS*

In the *Upaniṣads* speculations on language become more philosophical. We find in the *Upaniṣads* various descriptions on the nature of the highest and purest speech; Brahman is defined as the one reality, without a second, and is identified with speech.[14] *Māṇḍūkyopaniṣad* symbolizes the Real by the word *Aum*.[15] Brahman which is said to be speech is also said to be *Aum*,[16] just as leaves are held together by a stalk, so is all speech held together by *Aum*.[17]

Aum is discussed in a number of *Upaniṣads* and it forms the entire subject matter of the *Mundakopaniṣad*. *Aum* which is also called *Praṇava* represents the mystical aspect of *vāk*. In the *Upaniṣads* worship of *vāk* had been enjoined as a kind of spiritual practice that ultimately leads to a state of perpetual bliss.[18] So it was considered so powerful and worthy of reverence. According to Swami Nikhilananda, *Aum* is the *Nāda-brahman*, the sound Brahman, that is, reality expressed through sound.[19] This means that the real and the *Aum* are one, the symbol of the divine. It is

---

[13]R. PANIKAR, *Vedic Experience*, 89.

[14]Br. Up. 4.1.2.

[15]Ma. Up. 1.1.

[16]Tait. Up. 1.8.1.

[17]Ch. Up. 2.23.3.

[18]Ch. Up. 1.1.

[19]S. NIKHILANANDA, "Aum, the Word of Words," in *Language: An Inquiry into its Meaning and Function*, ed., R.N. ANSHEN (London: Kennikat Pres 1971), 82.

the primordial utterance which makes all utterances possible without getting identified with any. Our speech, therefore, is an unreal expression of the Real which is the highest speech or *Aum*.

3. *NITYA ŚABDA*: THE MĪMĀṂSAKĀ CONCEPT

The rules of vedic interpretations were formulated for the first time in India by Jaimini (c.25A.D.), the believed to be author of *Mīmāṃsā Sūtra* in a very systematic manner. The *Mīmāṃsā Sūtras* set out to examine the duty of human beings (*dharma*). Duty was defined as imperative. That which gives oneself some urge is called duty. *Śabda* (vedic word) is regarded as the only source of duty. It is a chief *pramāṇa* since it is the sole means for knowing injunctions and prohibitions. Again words are regarded as eternal because if they are momentary they cannot communicate meaning. Being the bearer of meaning they are eternal.[20] Hence the Mīmāṃsakas maintain *Veda* as eternal and impersonal sources of our duty and guiding forces of our life. Śabaraswamin, Kumārila Bhaṭṭa and Prabhākara are the three main exponents of the Mīmāṃsa system in later times. We shall discuss in some detail the different views held by these great masters on *śabda*.

A. ŚABARA ON *ŚABDA*

Śabara (c.400 A.D.) is the author of the famous work *Śabarabhāṣya*, the earliest existing commentary on *Jaimini Sūtras*. In fact it was this *bhāṣya* that became the basis for all later Mīmāṃsa works. Śabara defines word as a combination of a number of letters.[21] Jaimini and Śabara speak of four different classes of words, i.e., *nāma* (noun) *sarvanāma* (pronoun)

---

[20]R.C. PANDEYA, *Problem of Meaning,* 13-14.

[21]ŚABARA, *Śabarabhāṣya*, tr. G. JHA, (Baroda: Oriental Institute 1973), I. 1.5. Hereafter *Śabarabhāṣya*.

viśeṣaṇa (adjective) and *akhyāta* or the *karma-śabda* (verb).[22] Verbs possess the capacity to denote activity and are expressive of a transcendental potency, called *apūrva*.[23] Since the verb expresses activity (*karma*) and dynamic power (*apūrva*) Śabara considers verb as the main word in a sentence.[24] Śabara, further speaks of four characteristics of *śabda*. They are eternality (*nityatva*), non human origin (*apauruṣeyatva*), without parts (*niravayavatva*), incorruptible (*akṣaratva*) and the relationship between word and its meaning is inborn (*autpattika*).[25] Defining word as "a number of letters" Śabara rejects the *sphoṭa* theory of the Grammarians.

B. KUMĀRILA BHAṬṬA ON *ŚABDA*

Kumārila (c.650 A.D.) by commenting systematically on the *Śabarabhāṣya* has given a new direction to the Mīmāṃsa thought. He wrote his celebrated independent exposition on *Śabarabhāṣya* in three parts known as: *Slokavārtika, Tantravārttika and Tuptīka.* According to Kumārila a 'word' (*śabda*) is nothing more than a combination of letters. And its combination conveys the denotation. Quoting from the *Shāshṭtṛadīpikā* of Pārthasārathi, Jha says: "Just as the various subsidiary actions of an elaborate sacrifice combine to bring about the final result, so the various letters composing the word combine to accomplish the denotation of the meaning."[26]

---

[22]G.V. DEVASTHALI, *Mīmāṃsa The Vākyaśāstra of Ancient India* (Bombay: Book Sellers Publishing Co. 1959), 96-97.

[23]*Śabarabhāṣya*, IX. 1.1.

[24]*Śabarabhāṣya*, I. 1.4.

[25]G. OTMAR, *Hermenutics and Language in Pūrva Mīmāṃsā* (Delhi: Motilal Banarsidass 1983), 42-47.

[26]G. JHA, *The Prabhākara School of Pūrva Mīmāṃsa* (Delhi: Motilal Banarsidass 1978), 54. Hereafter G. JHA, *The Prabhākara School of Pūrva Mīmāṃsa.*

Kumārila asserts the natural capacity of words to express denotation. In his view words convey the word meaning and word meanings convey the mutual relations. This theory is known as "*abhihitānvaya*" theory.[27] He says further that even the individual letters should be considered as a *śabda* though they do not convey any meaning.[28] And for Kumārila the synthesis of the word-meaning makes the sentence-meaning.

C. PRABHĀKARA ON *ŚABDA*

Prabhākara (c.700 A.D.) believed to be the pupil of Kum ārila, is a reformer and innovator of the system of Kumārila. According to Prabhākara all sounds (words) heard are in the shape of some letter or the other, there being no *dhvani* or indistinct sound and a word cannot be regarded as some thing different from the letters which compose it.[29] Apart from the sentence each word has a definite meaning, though it functions to serve as part of a sentence. In contrast to Kumārila's view, Prabhākara considers both the individual word meanings and their mutual relation, as being conveyed by the words themselves. This is the theory known as "*anvitābhidhāna vāda.*"[30]

According to Prabhākara the 'meaning' is that which is expressed or denoted by the word. Meaning is understood through the sense perception. There is a certain potency in the letters themselves which brings about the apprehension of the thing denoted. This proves that the expressiveness of the word is something that belongs to it by its very nature. Therefore, word is held to be eternal and independent of human agency. It is for this reason that Prabhākara considers it to be the direct cause of

---

[27]K. RAJA, *Theories of Meaning,* 194.

[28]KUMĀRILABHAṬṬA, *Ślokavārtika*, tr., G. JHA (Delhi: Satguru Publications 1983), 262.

[29]G. JHA, *The Prabhākara School of Pūrva Mīmāṃsa*, 54.

[30]K. RAJA, *Theories of Meaning*, 193-194.

verbal cognition.[31] The meaning of words may be more explicitly known when they occur in an injunctive sentence. In other words, words are intelligible when they are related to something which has to be done.[32]

4. THE GRAMMARIANS' APPROACH TO *ŚABDA*

Patañjali, in the *Mahābhāṣya* speaks of various advantages resulting from the study of Grammar. First and foremost, Grammar is to be studied for the preservation of the *Vedas*.[33] Perhaps it was with this intention the grammatical analysis itself began for the first time. However, we are not concerned here with Grammar or grammatical analysis as such, rather we propose to give a brief account of the notion of *śabda* as has been envisaged by Yaska the first Etymologist and the great Grammarians Pāṇini and Patañjali.

A. ANALYSIS OF *ŚABDA* IN YASKA'S *NIRUKTA*

Unfortunately the time of Yaska is by no means yet certain. Scholars have different opinions. However, Lakshman Sarup, the editor of Yaska's *Nirukta* places it between 500 - 700 B.C.[34] *Nirukta* of Yaska is the oldest surviving Indian treatise on Etymology, Philosophy and Semantics. *Nirukta* mentions several authorities on the meaning and language known before Yaska. All of them are now extinct. That may be the reason why Lakshman considers Yaska to be the first writer on Etymology, in Indian tradition. Etymology according to Yaska is essential for the

---

[31]G. JHA, *The Prabhakara School of Pūrva Mīmāṃsa*, 62.

[32]G. JHA, *The Prabhākara School of Pūrva Mīmāṃsa*, 53-54.

[33]PATAÑJALI *Vyākaraṇa Mahābhāṣya*, tr., K.V. ABHANKAR & J.M. SUKLA (Poona: Bhandarkar Oriental Research Institute 1975), 7. Hereafter *Mahābhāṣya*.

[34]L. SARUP, *The Nighaṇṭu and the Nirukta* (Delhi: Motilal Banarsidass 1967), 53. Hereafter L. SARUP, *The Nighantu and the Nirukta*.

proper understanding of vedic texts. He considers it to be the complement of Grammar.[35] According to Belvalkar Yaska's work is a running commentary upon a list of words in five chapters, known as the *Nighaṇṭu* of which all the words are taken from the *Veda.*[36]

The *Nirukta* gives an analysis of the meaning of words. It is dealing mainly with meaning and thus it tries to explain the meaning of words on the basis of verbal roots. Yaska's fundamental notion about language is that all words can be reduced to their primordial elements which he calls roots (*dhātu*). Belvalkar observes it as a postulate of modern philosophy and the basis of the whole system of Pāṇini.[37] Yaska makes a fourfold classification of words: names, verbs, prefixes and preposition. He defines a 'name' as a word meaning a static referend and a verb as a word meaning a dynamic reality.[38] He also admits verbs to be the most important parts of language. However, Yaska has not made any attempt to define the word, *śabda.*

B. CONCEPT OF *ŚABDA* IN THE *AṢṬĀDHYĀYĪ* OF PĀṆINI

Pāṇini's (c.400 B.C.) *Aṣṭādhyāyī* is considered to be the oldest celebrated grammatical work in Sanskrit. The *Aṣṭādhyāyī* consists of approximately four thousand aphorisms divided into eight chapters. *Aṣṭādhyāyī* is also known as *Pāṇiniyam.*[39] Although no actual grammar has come down to us from the vedic period, the phonetic works called *Prātisākyas* and *Sikṣhas*, may be regarded

---

[35]L. SARUP, *The Nighaṇṭu and the Nirukta*, 14-15.

[36]S.K. BELVALKAR, *Systems of Sanskrit Grammar* (Varanasi: Bharatiya Vidya Prakashan 1976), 6. Hereafter S.K. BELVALKAR, *Systems of Sanskrit Grammar.*

[37]S.K. BELVALKAR, *Systems of Sanskrit Grammar*, 7.

[38]R.C. PANDEYA, *Problem of Meaning*, 12.

[39]B. WALKER, *Hindu World*, Vol. II (London: George Allen & Unwin 1968), 182.

as grammatical treatises.[40] The style (Algebraic and artificial)[41] of *Aṣṭādhyāyī* points to the presence of many such previous works. Prof. Keith says: "what is clear from Pāṇini's own works is that he summarizes the efforts of many previous writers, from whom we may be sure, he borrowed his form as well as many facts.[42]

Whatever be the case the analysis of language taken up by Paṇini is difficult to grasp because of its aphoristic style. There is no doubt that *Aṣṭādhyāyī* by its style, accuracy and thoroughness eclipsed all its predecessors.[43] Correct use of language was the prime concern of Pāṇini. So he was interested more with the form of language than with its meaning. Hence he has thoroughly investigated into the nature of the parts of speech and etymological derivations of words. He did not therefore enquire into the nature of relation between words and their referent or into the question of eternality of words.[44] As R.C. Pandeya points out, discussions in Katyayana's *Vārttika* and Patañjali's *Mahābhāṣya* make it clear that Pāṇini had the presupposition of the eternality of words.[45] According to K. Raja, Pāṇini himself mentioned this fundamental dichotomy between form and content in the following *sūtra*: *Svam rūpam śabdasya śabdasamjñā* - 'a word (in a grammatical rule) which is not a technical term denotes its own form'.[46]

A good many treatises and commentaries deal with grammatical works supplementary to Pāṇini. In this connection

---

[40]A. MAcDONELL, *India's Past, a survey of her Literatures, Religions, Languages and Antiquities* (Delhi: Motilal Banarsidass 1956), 140.

[41]S.K. BELWALKAR, *Systems of Sanskrit Grammar*, 7.

[42]A.B. KEITH, *A History of Sanskrit Literature* (London: Oxford University Press 1961), 423.

[43]S.K. BELWALKAR, *Systems of Sanskrit Grammar*, 10.

[44]R.C. PANDEYA, *Problem of Meaning*, 16.

[45]R.C. PANDEYA, *Problem of Meaning*, 17.

[46]K. RAJA, *Theories of Meaning*, 117.

the name of Pāṇini's successor Katyayana also should be mentioned. Katyayana, in his *Vārttika* criticizes the rules of Pāṇini with a view to correct and supplement.[47] Again Bhaṭṭoji's *Siddhāntakaumudi* is to be mentioned; for it is the most popular introduction to Pāṇini's Grammar. Patañjali also examines Pāṇini's *sūtras* in his *Mahābhāṣya.*

C. PATAÑJALI'S VIEWS ON *ŚABDA*

The *Mahābhāṣya* (great commentary) of Patañjali (150 B.C.) is an encyclopedia and authoritative work on Sanskrit Grammar. *Mahābhāṣya* is first and foremost a commentary on the *Pāṇini sūtras* and the various *Vārttikas* on it. He has commented upon only 1228 *sūtras* of Pāṇini out of the total number of 3972 according to the traditional recital and the number of *Varttikas* of these 1128 *sutras* goes above 5000. The very first statement of the *Mahābhāṣya*, "now commences the science of Grammar" - points to the fact that Patañjali has written his *Bhāṣya* as a scientific treatise on Sanskrit Grammar. Abhyankar observes three different branches of development in the science of Sanskrit grammar, founded by Patañjali. They are: formation of words, determination of sense and the connection between word and sense.[48]

Unlike Yaska and Pāṇini who were dealing with the word-element, Patañjali stresses on the sense element of a word. It is very clear from his definition of *śabda*. *Śabda* is defined by Patañjali as, "that by the utterance of which objects are cognised." An alternative definition of *śabda* is also proposed which states that, *śabda* is that sound which communicates objects.[49] It may be in this connection Raja speaks of Patañjali as distinguishing

---

[47] S.K. BELWALKER, *Systems of Sanskrit Grammar*, 24.

[48] *Mahābhāṣya*, iv.

[49] *Mahābhāṣya*, 5-6.

between the two aspects of a word, namely, the *sphoṭa* and the *dhvani*. The former is the permanent element in the word and may be considered the essential word, whereas the latter is the actualized and ephemeral element.[50]

There is also a discussion on the question whether word-meaning is universal or particular. And the Grammarian accepts both the possibilities.[51] It also speaks about the word, its sense and the connection between the two as already established.[52] Nevertheless, Patañjali did not consider the word as an invisible and timeless symbol, apart from the letters that are revealed when the word is uttered. But he was conscious of the meaning bearing aspects of word, though he did not explain it fully. According to him an intelligent man grasps the letters of the word together with the order of their sequence. So the unity of the word is based mainly on the unity of the meaning.[53] In this way, through his excellent analysis of *śabda* in the *Mahābhāṣya* Patañjali has marked the highest point in the development of the science of Grammar. This exposition provides us with a background for our study of the Grammarian Philosopher Bhartṛhari who has made the science of Grammar a science in search of the Real - a real philosophical pursuit.

---

[50]K. RAJA, *Theories of Meaning*, 100.

[51]*Mahābhāṣya*, 23.

[52]*Mahābhāṣya*, 24.

[53]K. RAJA, *Theories of Meaning*, 109.

BIBLIOGRAPHY

I. THE EDITED WORKS OF BHARTṚHARI

BHARTṚHARI. *Vākyapadīya,* Part I, ed. SASTRI, C. Lahore: Ram Lal Kapoor Trust 1935.

-----------. *Vākyapadīa*, Part III, ed. SASTRI, S. K. Trivandrum: Government Press 1935.

-----------. *Vākyapadīya*, With Commentary *Bhāvapradīpa*, ed. SUKLA, S. Varanasi: Sarasvati Bhavana Grandamala 1961.

-----------. *Vākyapadīya*, Parts I-III, ed. SARMA, R. Varanasi: Sarasvati Bhavana Granthamala 1963-1974.

-----------. *Vākyapadīyam*, Part I, ed. TRIPATHI, B. P. Varanasi: Sarasvati Bhavana Granthamala 1976.

-----------. *Vākyapadīyam*, Part III, ed. TRIPATHI, B. P. Varanasi: Sarasvati Bhavana Granthamala 1979.

-----------. *Vākyapadīyam,* Part II, ed. TRIPATHI, B. P. Varanasi: Sarasvati Bhavana Granthamala 1980.

I A. TRANSLATIONS, STUDIES AND OTHER RELATED WORKS

AKLUJAKAR, A. "The Concluding Verses of Bhartṛhari's *Vākya-kāṇḍa*." *Annals of the Bhandarkar Oriental Research Institute* 58/59 (1977-1978) 9-26.

-----------. "Ancient Indian Semantics." *Annals of Bhandarkar Oriental Research Institute* 51 (1970) 11-29.

-----------. "Nakamura on Bhartṛhari." *Indo Iranian Journal* 13/3 (1971) 161-175.

AMALADASS, A. *Philosophical Implications of Dhvani*. Vienna: Gerold & Co. 1984.

ĀNANDAVARDHANA. *Dhvanyāloka*, tr. KRISHNAMOORTY, K.

Poona: Oriental Book Agency 1955.

-------------. *Dhvanyāloka*, ed. SASTRI, P. Benares: Kashi Sanskrit Series 1940.

ANSHEN, R. N., ed. *Language: An Inquiry into its Meaning and Function*. London: Kennikat Press 1971.

AYENGAR, G. V. *Bhartṛhari's Nītiśatakam*. Thanjavur: T. M. S. M. Library 1980.

BELVALKAR, S. K. *Systems of Sanskrit Grammar*. Varanasi: Bharariya Vidya Prakashan 1976.

BHARTṚHARI. *Vākyapadīya, avec la Vṛtti de Harivṛsbha*, ed. BIARDEAU, M. Paris: E. De Boccard 1964.

-----------. *Vākyapadīya*, Chapter III, Part I, tr. IYER, K. A. S. Poona: Deccan College 1963.

----------. *Vākyapadīya*, Chapter III, Part II., tr. IYER, K. A. S. Poona: Deccan College 1974.

-----------. *Vākyapadīya*, ed. RAU, W. Wiesbaden: F. Steiner 1977.

-----------. *Vākyapadīya,* Parts I & II, Text & tr. PILLAI K. R. Delhi: Motilal Banarsidass 1971.

-----------. *Vākyapadīya*, Chapter I, tr. IYER, K. A. S. Poona: Deccan College 1965.

-----------. *Vākyapadīya*, Chapter II, tr. IYER, K. A. S. Delhi: Motilal Banarsidass 1977.

-----------. *Vākyapadīya, Brahmakāṇḍa,* tr. VARMA. Delhi: Munshiram Manoharlal 1970.

BHATE, S. & BRONKHORST, J., eds. *Bhartṛhari: Philosopher and Grammarian* (Proceedings of the first international conference on Bhartṛhari)Delhi: Motilal Banarsidass 1994.

BHAṬṬACHARYA, B. *Bhartṛhari's Vākyapadīya and Linguistic Monism*. Poona: Bhandarkar Oriental Research Institute 1985.

BHAṬṬACHARYA, J. *et al*., tr. *Nyāyamañjarī* in *Encyclopedia of Indian Philosophies* Vol. II, The Tradition of Nyāya-Vaiśeṣika up to Gangeśa, ed. K.H. POTTER. Delhi: Motilal Banarsidas 1978.

BHAṬṬOJI DĪKSITA. *Śabdakaustubha.* Benares: Chowkhamba Sanskrit Series 1933.

BLOOMFIELD, L. *Language*. Delhi: Motilal Banarsidass 1994.

CARPENTER, D. *Revelation, History, and the Dialogue of Religions*. New York: Orbis Books 1995.

CHAKRABARTI, A. "Sentence-Holism, Context-Principle and Connected-Designation *Anvitabhidhāna*: Three Doctrines or One?" *Journal of Indian Philosophy* 17 (1989) 37-41.

CHAKRAVARTI, P. C. *The Linguistic Speculations of the Hindus*. Calcutta: Calcutta Uuniversity Press 1924.

------------. *Philosophy of Sanskrit Grammar*. Calcutta: University of Calcutta 1930.

CHATTOPADHYAYA, D. P. *et al. Phenomenology and Indian Philosophy*. Delhi: Motilal Banarsidass 1992.

CHRISTIE, E. "Indian Philosophers on Poetic Imagination (*Pratibhā*)." *Journal of Indian Philosophy* 77/ 2 (1979) 153-203.

COWARD, H. G. & RAJA, K., eds. *Encyclopedia of Indian Philosophies: The Philosophy of the Grammarians*. Vol. V. Delhi: Motilal Banarsidass 1990.

COWARD, H. G. The *Sphoṭa* Theory of Language. Delhi: Motilal Banarsidass 1980.

------------. "Derrida and Bhartṛhari's *Vākyapadīya* on the Origin of Language." *Philosophy East West* 40/1 (1990) 3-16.

------------. "Scripture, Revelation and Consciousness in Christianity and Hinduism." *Journal of Dharma* 3/3 (1978) 238-252.

------------. *Derrida and Indian Philosophy*. Albany: State University of NY Press 1990.

-------------. *Bhartṛhari*. Boston: Twayne Publications 1976.

------------. "The Yoga of the Word (*Śabdapūrvayoga*)." *Adyer Library Bulletin* 49 (1985) 1-13.

------------. "Language as Revelation." *Indian Philosophical Quarterly* 3 (1976) 447-472.

------------. "Time (*kāla*) in Bhartṛhari's *Vākyapadīya*." *Journal of Indian philosophy* 10/3 (1982) 277-287.

DASGUPTA, S. *Indian Idealism*. Cambridge University Press 1962.

DEUTSCH, E., ed. *Culture and Modernity: East West Philosophic Perspectives*. Delhi: Motilal Banarsidass 1994.

DEVARJ, N. K. "Language, Logic and Reality." *Indian Philosophical Quarterly* 17/3 (1990) 325-347.

DEVASTALI, G. V. *Mīmāṃsa: The Vākyaśāstra of Ancient India.* Bombay: Booksellers Publishing Company 1959.

DRAVID, R. R *The Problem of Universals in Indian Philosophy.* Delhi: Motilal Banarsidass 1980.

FRAUWALLNER, E. *History of Indian Philosophy.* Delhi: Motilal Banarsidass 1984.

GANERI, J. "Meaning and Reference in Classical India." *Journal of Indian Philosophy* 24 (1996) 1-19.

GANGULI, H. K. *Philosophy of Logical Construction.* Calcutta: Sanskrit Pustak Bhandar 1963.

GOKHALE, P. P. "Language and World: Some Classical Indian Approaches vis-a-vis Analytical Western Approaches." *Indian Philosophical Quarterly* 21/9 (1994) 317-328.

GOPINATH, P. The *Nīti Śataka, Śriṇgāra Śataka* and *Vairāgya Śataka* of Bhartṛhari. Bombay: Khemraj shri krishnadass 1914.

GRIFFITH, R. T. H., tr. *The Hymns of Rigveda,* Vols. I & II. Varanasi: Chowkhamba Sanskrit Series 1963.

HERZBERGER, H. G. & HERZBERGER, R. "Bhartṛhari's Paradox." *Journal of Indian Philosophy* 9 (1981) 1-18.

HOUBEN, J. E. M. *The Sambandha Samuddeśa (Chapter on Relation) and Bhartṛhari's Philosophy of Language.* Delhi: Motilal Banarsidass 1995.

------------. "Bhartṛhari's *Samaya* / Helārāja's *Saṃketa.*" *Journal of Indian Philosophy* 20 (1992) 219-242.

INGALLS, D. H. H. "The Study of Śaṇkaraṃkarācārya." *Annals of Bhandarkar Oriental Research Institute* 33 (1952) 1-14.

------------. "The Comparison of Indian and Western Philosophy." *Journal of Oriental Research* 22 (1954).

IYER, K. A. S. "The Doctrine of *Pratibhā* in Indian Philosophy." *Annals of Bhandarkar Oriental Research Institute* 5 (1953) 1-18;113-132.

------------. *Bhartṛhari: A study of the Vākyapadīya in the Light of Ancient Commentaries.* Poona: Deccan College 1969.

------------. "Bhartṛhari on Dhvani." *Annals of the Bhandarkar Oriental Research Institute* 46 (1965) 49-65.

------------. "Bhartṛhari on *Apabhraṃśa.*" *Vishveshvarand Indological*

Journal 2 (1964) 242-246.

------------. *The Vākyapadīya: Some problems*. Poona: Bhandarkar Oriental Research Institute 1982.

------------. "Bhartṛhari on the Primary and Secondary Meanings of Words." *Indian Linguistics* 29 (1968) 97-112.

------------. "Bhartṛhari on *Vyākaraṇa* as a means of attaining *Mokṣa*." *Adayar Library Bulletin* 28 (1964) 112-131.

JAIMINI. *Mīmāṃsa Sūtra*, tr. SANDAL, M. Delhi: Motilal Banarsidass 1980.

JHA, G. *Prabhākara* School of Pūrva-Mīmāṃsa. Delhi: Motilal Banarsidass 1978.

JAYANTABHAṬṬA. *Nyāyamañjarī*, ed. SUKLA, S. N. Benaras: Jaya krishna Das Haridas Gupta 1936.

JOSHI, S. D., ed. *Proceedings of the Winter Institute on Ancient Indian Theories on Sentence -meaning*. Poona: University of Poona 1979.

KAPIL, D. S. "Bhartṛhari's Discussion on the Relation between *Upāmana* and *Upameya* in his *Vākyapadīya*." *Indian Linguistics* 25 (1964) 229-233.

KAVIRAJ, G. "The Doctrine of *Pratibhā* in Indian Philosophy." *Annals of Bhandarkar Oriental Research Institute* 5 (1923-24) 1-18.

KEITH, B. A. *A History of Sanskrit Literature*. London: Oxford University Press 1961.

KOCMARCK, I. *Language and Release (Saravjñātman's Pañcaprakriyā)*. Delhi: Motilal Banarsidass 1985.

KUMAR, P. *Linguistic Thought in Ancient India*. Delhi: Nag publishers 1984.

KUMĀRILABHAṬṬA. *Ślokavārtika*, tr. JHA, G. Delhi: Sri Satguru Publications 1983.

LUKE, K. "Some Aspects of the Rigvedic Conception of *Vāk*." *Jeevadhara* 1/2 (1971) 219-236.

MADHAVA. *Sarva Darśana Saṃgraha*, tr. COWELL, E. B. & GOUGH, A. E. Delhi: Cosmo Publications 1976.

MANDANA MISRA. *Sphoṭasiddhi*, tr. IYER, K. A. S. Poona: Deccan College 1966.

MANICKAM, T. "Cross Cultural Hermeneutics: the Patterns of Jaimini, Bhartṛhari and Śaṇkarāchārya." *Theological Studies* 21/3-4 (1984) 250-267.

------------. "Toward an Indian Hermeneutics of the Bible." *Jeevadhara* 12/68 (1982) 94-105.

------------, ed. *Seminar on Vedic Scriptures*. California: UCSB 1978.

MANNINEZHATH, T. "Advaita Critique of the *Sphoṭa* and *Śabdabrahman*." *Journal of Dharma* 17/3 (1992) 178-195.

MASATOSHI, N. *Sanskrit and Indian Studies*. London: D. Reidel Publishing Company 1980.

MATILAL, B. K. *Logic, Language and Reality*. Delhi: Motilal Banarsidass 1990.

------------. *The Word and the World: India's Contribution to the Study of Language*. Delhi: Oxford University Press 1990.

MATILAL, B. K. & SHAW, J. L. *Analytical Philosophy in Comparative Perspective*. Dordrecht: Reidel 1985.

McDONNELL, A. *India's Past, A survey of Her Literatures, Religions, Languages, and Antiquities*. Delhi: Motilal Banarsidass 1956.

MOORTY, K. K. "Observations of Sanskrit Literary Critics on Poetic Imagination." *Poona Orientalist* 9 (1944) 123-132.

------------. "Sanskrit Theories of Poetry." *Poona Orientalist* 8 (1943) 9-20.

MÜLLER, M. F., ed. *Sacred Books of the East*. Vol. XXXIV: *Vedānta Sūtras with the Commentary by Śankarācārya*, Part I, tr. THIBAUT, G. Delhi: Motilal Banarsidass 1973.

MURTI, T. R. V. "Some Comments on the Philosophy of Language in the Indian Context." *Journal of Indian Philosophy* 2 (1974) 321-331.

MURTI, M. S. "Bhartṛhari on *Svaṃ rūpam Śabdasyaśabdasaṃjñā*." *Adyer Libray Bulletin* 44/45 (1980) 602-613.

NĀGEŚABHAṬṬA. *Sphoṭavāda*, ed. KRISHNAMACHARYA, V. Madras: The Adyar Library 1946.

NAKAMURA, H. "The Concept of Brahman in Bhartṛhari's Philosophy." *Journal of Oriental Research* 40/41 (1981) 135-149.

NANDI, S. K. *Studies in Modern Indian Aesthetics*. Simla: Indian Institutes of Advanced Study 1975.

OTHMAR, G. *Hermeneutics and Language in Pūrva-Mīmāṃsā*. Delhi: Motilal Banarsidass 1983.

PAL, S. K. "The Existential Perspective of Language and Meaning." *Darśana International* 32 (1992) 33-41.

PANDEY, K. C. *Comparative Aesthetics*. Vol. I. Varanasi: Chowkhamba Sanskrit Series 1959.

PANDEY, S. L. *Pre-Śaṇkara Advaita Philosophy*. Allahabad: Darshan Peeth 1974.

PANDEYA, R. C. *The Problem of Meaning in Indian Philosophy*. Delhi: Motilal Banarsidass 1963.

PANIKKAR, R. *The Vedic Experience*. London: Darton, Longman & Todd 1979.

PANSIKAR, W. L. S., ed. *Subhāshita-Trisati of Bhartṛhari* Bombay: Pandurang Jawaji 1925.

PATAÑJALI. *Vyākaraṇa Mahābhāṣya*, tr. ABHYANKAR, K. V. & SHUKLA, J. M. Poona: Bhandarkar Oriental Institute 1975.

PATHIRAJ, R. "The *Sphoṭa* Doctrine of Bhartṛhari." *Indian Philosophical Quarterly* 22/1 (1995) 67-74.

PATNAIK, T. "Language, Thought and Communication: An Appraisal of Bhartṛhari's Theory of Language." *Indian Philosophical Quarterly* 15/3 (1988) 349-368.

PLOTT, J. C. *Global History of Philosophy*. Vol. III. Delhi: Motilal Banarsidass 1980.

POTTER, K. H., ed. *Encyclopedia of Indian Philosophies: Advaita Vedanta up to Śamkara and His Pupils*. Vol. III. Delhi: Motilal Banarsidass 1981.

PRAHARAJ, D. M. "Speech-Acts: As Linguistic Communicative Function." *Indian Philosophical Quarterly* 22 (1995) 225-237.

RADHAKRISHNAN, S. *The Principal Upaniṣads*. London: George Allen &Unwin Ltd. 1953.

RAJA, K. "Bhartṛhari's List of Sentence-definitions." *Adyar Library Bulletin* 26 (1962) 206-210.

-----------. *Indian Theories of Meaning*. Madras: Adyar Library and Research Centre 1977.

RAJU, P. T. *Introduction to Comparative Philosophy*. Delhi: Motilal Banarsidass 1992.

------------. *Structural depths of Indian Thought*. New Delhi: South Asian Publishers 1985.

------------. "Idealisms: Eastern and Western." *Philosophy East and West* 5/3 (1955) 211-234.

ROEST CROLLIUS, A. A. *Thus Were They Hearing*. The Word in the experience of Revelation in Qur'ān and Hindu Scriptures. Roma: Università Gregoriana 1974.

ŚABARA. *Śabarabhāṣya*, tr. JHA, G. Baroda: Oriental Institute 1973.

SARMA, E. R. S. "Bhartṛhari as an Advaitin." *Adyar Library Bulletin* 50 (1986) 579-585.

SARMA, M. K. "Bhartṛhari: A Great Post Upaniṣhadic Intuitionist." *The Aryan Path* 11 (1940) 539-549.

------------. "*Vāk* Before Bhartṛhari." *Poona Orientalist* 8 (1943) 21-36.

SHARMA, M. M. *The Dhvani Theory in Sanskrit Poetics*. Varanasi: Chowkhamba Publications 1968.

SHARMA, P. S. "The Idea of Power (*śakti*) in the *Vākyapadīya*." *Bhartiya vidya* 42/1-4 (1982) 1-21.

------------. *The Kālasamuddeśa of Bhartṛhari's Vākyapadīya*. Delhi: Motilal Banarsidass 1972.

SHARMA, V. "Is the Eternal Verbum of Bhartṛhari a Dynamic Principle?" *Vishveshvarand Indological Journal* 13 (1975) 337-350.

SARUP, L. *The Nighantu and the Nirukta*. Delhi: Motilal Banarsidass 1967.

SASTRI, S. *Essays on Indology*. Delhi: Mecarchand Lachmandass 1963.

SASTRI, G. N. *Elements of Indian Aesthetics*. Varanasi: Chawkhambha Orientalia 1978.

SASTRI, V. A. R. "*Mukhya* and *Gauṇa* words in Language." *Bulletin of the Deccan College Research Institute* 14/3 (1952) 183-193.

------------. "Bhartṛhari as a *Mīmāṃsaka*." *Bulletin of the Deccan College Research Institute* 14/1 (1952) 1-15.

SASTRI, G. *The Philosophy of Word and Meaning*. Calcutta: Sanskrit College 1959.

----------. "The Doctrine of *Śabdabrahman* - a criticism by Jayanthabhaṭṭa." *Indian Historical Quarterly* 15/3 (1939) 441-453.

-----------. A study in the Dialectics of *Sphoṭa*. Delhi: Motilal Banarsidass 1980.

SIDERITS, M. *Indian Philosophy of Language*. Dordrecht: Kluwer Academic Publishers 1991.

-----------. "Word meaning, Sentence meaning, and *Apoha*." *Journal of Indian Philosophy* 13 (1995) 133-151.

SREEKANTAIYA, T. N. "Imagination in Indian Poetics." *The Indian Historical Quarterly* 13(1937) 59-84.

STAAL, J. F. *A Reader on the Sanskrit Grammarians*. Delhi: Motilal Banarsidass 1985.

STUTLEY, J. & MARGERET. *Dictionary of Hinduism*. New York: Harper and Row 1977.

SUBBARAO, V. *The Philosophy of Sentence and Its Parts*. New Delhi: Munshiram Manoharlal 1969.

TABER, J. A. "The Theory of the Sentence in Pūrva- Mīmāṃsa and Western Philosophy." *Journal of Indian Philosophy* 17 (1989) 407-430.

TIWARI, D. N. "Bhartṛhari's Philosophy of Relation Between Word and Meaning." *Journal of Indian Council of Philosophical Research* 11/2 (1994) 43-54.

TOLA, F. & DRAGONETTI, C. "Some Remarks on Bhartṛhari's Concept of *Pratibhā*." *Journal of Indian Philosophy* 18 (1990) 95-112.

TORELLA, R. "Examples of the Influence of Sanskrit Grammar on Indian Philosophy." *East and West* 37 (1987) 151-164.

VARMA, S. K. "Definition of Sentence." *International Journal of Dravidian Linguistics* 13 (1984) 171-176.

VARMA, S. *Studies in Indology*. Delhi: Bhartiya Prakashan 1981.

VRAT, S. "The Concept of Time according to Bhartṛhari." *Annals of Bhandarkar Oriental Research Institute* 39 (1958) 68-78.

WALKER, B. *Hindu World.* Vols. II. London: George Allen & Unwin Ltd. 1968.

II. WORKS BY MARTIN HEIDEGGER

HEIDEGGER, M. *Sein und Zeit* (1927). GA. 2. Frankfurt am Main: Vittorio Klostermann 1977.

------------. *Die Grundprobleme der Phänomenologie* (1927). GA. 24. Frankfurt am Main: Vittorio Klostermann 1975.

------------. *Kant und Das Problem Der Metaphysik* (1929). GA. 3. Frankfurt am Main: Vittorio Klostermann 1991.

------------.*Vom Wesen des Grundes* (1929). In *Wegmarken*, GA. 9. Frankfurt am Main: Vittorio Klostermann, 1976.

------------.*Was ist Metaphysik*? (1929). In *Wegmarken*, GA. 9. Frankfurt am Main: Vittorio Klostermann 1976.

------------. *Vom Wesen der Wahrheit* (1930). In *Wegmarken*, GA. 9. Frankfurt am Main: Vittorio Klostermann 1976.

------------. *Einführung in die Metaphysik* (1935). GA. 40. Frankfurt am Main: Vittorio Klostermann 1983.

------------. *Erläuterungen zu Hölderlins Dichtung* (1936). GA 4. Frankfurt am Main: Vittorio Klostermann 1981.

------------. *Vorträge und Aufsätze* (1936-1954). Pfullingen: Neske 1954.

------------. *Beiträge zur Philosophie (vom Ereignis)* (1936-1938). GA. 65. Frankfurt am Main: Vittorio Klostermann 1989.

------------. *Nachwort zu "Was ist Metaphysik?"* (1943). In *Wegmarken*, GA. 9. Frankfurt am Main: Vittorio Klostermann 1976.

------------. *Brief über den "Humanismus"*(1946). In *Wegmarken*, GA. 9. Frankfurt am Main: Vittorio Klostermann 1976.

------------. *Einleitung zu "Was ist Metaphysik?"* (1949). In *Wegmarken*, GA. 9. Frankfurt am Main: Vittorio Klostermann 1976.

------------. *Holzwege*. Frankfurt am Main: Vittorio Klostermann 1950.

------------. *Unterwegs zur Sprache* (1950-1959). GA. 12. Pfullingen: Neske 1985.

------------. *Was heißt Denken?* (1951-1952). Tübingen: Niemeyer 1961.

------------. *Aus der Erfahrung des Denkens* (1954). Pfulligen: Neske 1954.

------------. *Was ist das - die Philosophie*? (1955). Pfullingen: Neske 1956.

------------. *Zur Seinsfrage* (1955). In *Wegmarken*, GA. 9. Frankfurt am Main: Vittorio Klostermann 1976.

------------. *Der Satz Vom Grund* (1957). Pfullingen: Neske 1957.

------------. *Identität und Differenz* (1957). Pfullingen: Neske 1957.

------------. *Gelassenheit* (1959). Pfullingen: Neske 1977.

------------. *Zeit und Sein* (1962). In *Zur Sache Des Denkens*. Tübingen: Niemeyer 1969.

II. A. TRANSLATIONS, STUDIES AND OTHER RELATED WORKS

ADAMCZEWSKI, Z. "Martin Heidegger and Man's Way to be." *Man and World* I (1968) 363-379.

ADKINS, A. W. "Heidegger and Language." *Philosophy* 37 (1962) 229-237.

ALSTON, W. P. & NAKHNIKIAN, G., eds. *Readings in Twentieth - Century Philosophy.* London: Collier-Macmillan Ltd 1963.

AMBROSIO, F. J. "Dawn and Dusk: Gadamer and Heidegger on Truth." *Man and World* 19 (1986) 21-53.

ANDERSON, S. W. "Dasein as Self: Some Implications of Heideggerian Ontology." *Kinesis* 14 (1985) 106- 129.

ARISAKA, Y. "Spatiality, Temporality, and the Problem of Foundation in Being and Time." *Philosophy Today* 40/1 (1996) 36-46.

ARLEEN, B. D. *et al.*, eds. *Crises in Continental Philosophy.* Albany: State University of NY Press 1990.

BAILIFF, J. D. *Coming to be: An Introduction of the Self in the Thought of Martin Heidegger.* Ann Arbor: University of Microfilms 1984.

BALLARD, E. G. & SCOTT, C. E., eds. *Martin Heidegger in Europe and America.* The Hague: Martinus Nijhoff 1973.

BARASH, J. A. *Martin Heidegger and the Problem of Historical Meaning*. Dordrecht: Martinus Nijhoff 1988.

BARRETT, W. & ALKEN, H. D., eds. *Philosophy in the Twentieth Century*. Vol. II. New York: Random House 1962.

BARTKY, S. "Originative Thinking in the Later Philosophy of Heidegger." *Philosophy and Phenomenological Research* 30 (1969) 368-381.

BERNASCONI, R. *The Question of Language in Heidegger's History of Being*. Atlantic Highlands, NJ: Humanities Press International 1989.

BLACHHAM, H. J., ed. *Reality, Man and Existence: Existential Works of Existentialism.* New York: Bantham Books 1965.

BLACKSON, T. A. "Plato and the Senses of Words." *Journal of the History of Philosophy* 29 (1991) 169-182.

BONSOR, J. A. *Rahner, Heidegger and Truth: Karl Rahner's Notion of Christian Truth, the Influence of Heidegger*. Lanham: University Press of America 1987.

BORGMANN, A. "The Transformation of Heidegger's Thought." *The Personalist* 47 (1966) 484-499.

BROCK, W., ed. *Existence and Being*. Washington: Regnery Gateway 1988.

BRUNS, G. L. *Heidegger's Estrangements: Language, Truth, and Poetry in the Later Writings*. New Haven: Yale University Press 1989.

BRUZINA, R. & WILSHIRE, B., eds. *Phenomenology Dialogues and Bridges*. Albany: State University of NY Press 1982.

BUCHANA, J. H. "Heidegger and the Problem of Ground: An Evaluation." *Philosophy Today* 17 (1973) 232-243.

CAPUTO, J. D. "Heidegger and Aquinas: The Thought of Being and Metaphysics of Esse." *Philosophy Today* 26 (1982) 194-203.

------------. *The Mystical Element in Heidegger's Thought*. Ohio: University of Ohio Press 1978.

------------. "Heidegger's God and the Lord of the History." *New Scholasticism* 47/4 (1983) 439-464.

-----------. *Heidegger and Aquinas: An Essay on Overcoming Metaphysics*. New York: Fordham University Press 1982.

CERF, W. "An Approach to Heidegger's Ontology." *Philosophy and Phenomenologic Research* 1 (1940) 177-190.

CHATTOPADHYAYA, D. P. *Knowledge, Freedom and Language*. Delhi: Motilal Banarsidass 1989.

CHENG, C. Y. "Remarks on the Ontological and Transontological Foundations of Language." *Journal of Chinese Studies* 5 (1978) 335-340.

COMAY, R. "After Metaphysics: On the Way to Heidegger." *Man and World* 19 (1986) 225-240.

COUSINEAU, R. H. *Humanism and Ethics: An Introduction to Heidegger's "Letter on Humanism" with a Critical Bibliography*. Paris: Nauwelaerts 1972.

CROWNFIELD, D. R. "The Question of God Thinking after Heidegger." *Philosophy Today* 40/1 (1996) 47-54.

DAVID, W. *Heidegger and Language: A Collection of Original Papers*. Coventry: University of Warwick 1981.

DEELY, J. N. *The Tradition via Heidegger: An Essay on the Meaning of Being in the Philosophy of Martin Heidegger*. The Hague: Martinus Nijhoff 1971.

DEMSKE, J. M. *Being, Man and Death: A Key to Heidegger*. Kentucky: The University Press 1970.

------------. "Heidegger's Quadrate and Revelation of Being." *Philosophy Today* 7 (1963) 159-244.

DEUTSCH, E. "Commentary on J. C. Mehta's "Heidegger and the Comparison of Indian and Western Philosophy."" *Philosophy East and West* 20 (1970) 319-321.

DOSTAL, R. J. "Beyond Being: Heidegger's Plato." *Journal of the History of Philosophy* 23/1 (1985) 71-98.

DOUGLAS, G. H. "Heidegger's Notion of Poetic Truth." *The Personalist* 47 (1966) 500-508.

DREYFUS, H. L. & HALL, H., eds. *Heidegger: A Critical Reader*. Oxford: Blackwell 1992.

DREYFUS, H. L. *Being-in-the-World: A Commentary on Heidegger's Being and Time*. Cambridge, Massachusetts: The MIT Press 1991.

DRISCOLL, G. "Heidegger's Ethical Monism." *The New Scholasticism* 42 (1968) 561-577.

EMAD, P. "Reference, Sign and Language: Being and Time, Section

17." *The Collegium Phenomenologicum*: The first Ten Years. Dordrecht: D. Reidel Publishing Co. (1988) 175-189.

------------. "The Significance of the New Edition of *Subjekt* und *Dasein* and the Fundamental Ontology of Language." *Heidegger Studies* 2 (1986) 141-151.

ERICKSON, S. A. "Meaning and Language." *Man and world* 1/4 (1968) 536-586.

FABRO, C. "The Problem of Being and the Destiny of Man." *International Philosophical Quarterly* 1 (1961) 407-436.

FARBER, M. "Heidegger on the Essence of Truth." *Philosophy and Phenomenological Research* 18 (1958) 523-532.

FAY, T. A. "Heidegger: Thinking as Noein." *The Modern Schoolman* 51 (1973) 17-28.

------------. "Heidegger on Logic and Language. Some aporiai." *Tijdschrift Voor Filosofie* 52/3 (1990) 421-442

FEHÈR, I. M. "Heidegger's Postwar Turn." *Philosophy Today* 40/1 (1996) 9-35.

FRENCH, P. A. *et al. Contemporary Perspectives in the Philosophy of Language* II. Notre Dame: University of Notre Dame Press 1989.

FRINGS, M. S., ed. *Heidegger and the Quest for Truth.* Chicago: Quadrangle Books 1968.

GADAMER, H. G. "The Problem of Being in Heidegger and the Scholastics." *The Thomist* 41 (1977) 62-91.

------------. *Truth and Method.* London: Sheed & Ward 1975.

GARY, G. J. "Heidegger's Being." *The Journal of Philosophy* 49 (1952) 415-422.

GASCHÈ, R. "Perhaps - a Modality? On the way with Heidegger to Language." *Graduate Faculty Philosophy Journal* 2 (1993) 467-484.

GELVEN, M. *A Commentary on Heidegger's "Being and Time": A Section by Section Interpretation.* New York: Harper and Row 1970.

GELVEN, M. "Language as Saying and Showing." *Journal of Value Inquiry* 17 (1983) 151-163.

GERBER, R. J. "Heidegger: Thinking and Thanking Being." *The*

Modern Schoolman 44 (1967) 205-222.

GHANOTAKIS, G. "Heidegger's Fundamental Ontology." *Man and World* 14 (1981) 251-264.

GOEDECKE, R. "Linguistic Ontology." *Philosophy Today* 20 (1976) 157-166.

GOFF, R. "Saying and Being with Heidegger and Parmenides." *Man and World* 5 (1972) 62-78.

GOULD, C. C. *Authenticity and Being-with-Others: A Critique of Heidegger's Sein und Zeit*. London: University of Microfilms International 1977.

GUIGNON, C. B., ed. *The Cambridge Companion to Heidegger*. Cambridge: Cambridge University Press 1993.

------------. "Truth as Disclosure: Art, Language & History." *The Southern Journal of Philosophy* 28 (1990) 105-120.

------------. *Heidegger and the Problem of Knowledge.* Indianapolis: Hacket Publishing Co. 1983.

HAAR, M. *Heidegger and the essence of Man*. New York : State University of New York Press 1993.

HARRIES, K. "Wittgenstein and Heidegger: The Relationship of the Philosopher to Language." *Journal of Value Inquiry* 2 (1968-69) 281-291.

------------. "Heidegger's Conception of the Holy." *The Personalist* 47 (1966) 169-184.

HART, J. G. & MARALDO, J. C., ed. *The Piety of Thinking*. Indiana University Press 1976.

HASS, L. "Dasein and Others: Heidegger's Ontology of Intersubjectivity." *Auslegung* 15 (1989) 81-93.

HEIDEGGER, M. *An Introduction to Metaphysics.* tr. MANHEIM, R. New Haven: Yale University Press 1959.

------------. *The Basic Problems of Phenomenology*. tr. HOFSTADTER, A. Bloomington: Indiana University Press 1982.

------------. "The way back into the ground of Metaphysics." tr. W. KAUFMANN in *Existentialism from Dostoevsky to Sartre*. ed. KAUFMANN, W. NY: World-Meridian 1956.

------------. *Identity and Difference*. tr. STAMBAUGH, J. New York:

Harper & Row 1974

------------. *What is Metaphysics*. tr. HULL R.F.C. & CRICK, A. in BROCK, W. *Existence and Being*. Washington: Regnery Gateway 1988.

------------. "Only a God Can save Us: *Der Spiegel*'s Interview with Martin Heidegger." tr. ALTER, M. P. & CAPUTO, J. D. *Philosophy Today* 20 (1976) 267-284.

------------. *What Is Called Thinking?* tr. WIECK F.D. & GRAY, J.G. New York: Harper & Row 1968.

------------. *Kant and the Problem of Metaphysics*. tr. CHURCHILL, J. S. Bloomington: Indiana University Press 1965.

------------. *The Question of Being*. tr. KLUBACK, W. & WILDE, J. T. New Haven: College and University Press 1958.

------------. *The Essence of Reasons*. tr. MALICK, T. Evanston: Northwestern University Press 1969.

------------. *On Time and Being*. tr. STAMBAUGH, J. New York: Harper and Row 1972.

------------. *The End of Philosophy*. tr. STAMBAUGH, J. New York: Harper and Row 1973.

------------. *What is Philosophy*. tr. KLUBACK, W. & WILDE, J. T. Albany: NCUP, Inc. 1995.

------------. Discourse on Thinking. tr. ANDERSON, J. M. & FREUND, H. New York: Harper Torch Books 1996.

------------. *The Concept of Time*. tr. McNEILL, W. Oxford: Blackwell Publishers 1996.

------------. *Poetry, Language and Thought*. tr. HOFSTADTER, A. New York. Harper and Row 1971.

------------. *Thoughts*. tr. HOELLER, K. *Philosophy Today* 20 (1976) 286-291.

------------. *The Principle of Ground*. tr. HOELLER, K. *Man and World* 4 (1971) 207-222.

------------. *Being and Time*. tr. MACQUARRIE, J. & ROBINSON, E. New York: Harper Collins 1962.

-----------. *On the Way to Language*. tr. HERTZ, P. D. New York: Harper and Row 1971.

-----------. *Early Greek Thinking*. tr. KRELL, D. F.& CAPUZZI, F. A.

San Francisco: Harper and Row 1975.

HEIM, M. "The Finite Framework of Language." *Philosophy Today* 31 (1987) 3-20.

HEINE, S. *Existential and Ontological Dimensions of Time in Heidegger and Dogen*. Albany: Sunny Press 1985.

HERMAN, R. *Heidegger and Derrida. Reflections on Time and Language.* Lincoln: University of Nebraska Press 1989.

HERRMANN, F. W. "The Flower of the Mouth": Hölderlin's hint for Heidegger's thinking of the essence of Language." *Research in Phenomenology* 19 (1989).

HIRSCH, E. F. "Martin Heidegger and the East." *Philosophy East West* 20 (1970) 247-263.

HOELLER, K. "The Role of the Early Greeks in Heidegger's Turning." *Philosophy Today* 28 (1984) 44-51.

HOLMES, R. H. "The World According to Husserl and Heidegger." *Man and World* 18 (1985) 373-387.

IHDE, D. "Phenomenology and the Later Heidegger." *Philosophy Today*18 (1974) 19-31.

IJSSELING, S. "Heidegger and the Destruction of Ontology." *Man and World* 15 (1982) 3-16.

JAMES, E. C. *The Authority of Language: Heidegger, Wittgenstein, and the Threat of Philosophical Nihilism.* Tampa: University of South Florida Press 1990.

JAMES, L. P. *Heidegger and the Divine: The Thinker, The Poet and God*. Athens: Ohio University Press 1974.

JONAS, H. "Heidegger and Theology." *The Review of Metaphysics* 18 (1964) 207-233.

KAHN, C.H. *The Verb 'Be' in Ancient Greek*. Dordrecht: Reidel 1973

KAUFMANN, W., ed. *Existentialism from Dostoevsky to Satre*. NY: World-Meridian 1956.

KELLY, T. A. *Language and Transcendence. A Study in the Philosophy of Martin Heidegger and Karl Otto Apel.* New York: European University Studies 1994.

KEVIN, M., ed. *Language, Truth and Ontology*. Dordrecht: Kluwer 1992.

KING, M. *Heidegger's Philosophy: A Guide to his Basic Thought.*

Oxford: Basil Blackwell 1964.

KISIEL, T. & BUREN, J. V., eds. *Reading Heidegger from the Start: Essays in His earliest Thought*. Albany: State University of New York Press 1994.

KISIEL, T. *The Genesis of Heidegger's Being and Time*. Berkeley: University of California Press 1993.

KOCKELMANS, J. J., ed. *On Heidegger and Language*. Evanston: Northwestern University Press 1972.

------------., ed. *A Companion of Martin Heidegger's "Being and Time."* Lanham: University Press of America 1986.

------------. *On the Truth of Being: Reflections on Heidegger's Later Philosophy*. Bloomington: Indiana university Press 1984.

KOVACS, G. *The Later Heidegger and Theology*. New York: Harper 1963.

------------. "The Leap (der Sprung) for Being in Heidegger's *Beiträge zur Philosophie* (vom Ereignis)." *Man and World* 25 (1992) 39-59.

------------. "The Ontological Difference in Heidegger's *Grundbegriffe*." *Heidegger Studies* 3/4 (1987/1988) 61-74.

------------. *The Question of God in Heidegger's Phenomenology*. Evanston: Northwestern University Press 1990.

KRELL, D. F., ed. *Basic Writings*. New York: Harper Collins 1993.

KURIAN, T. K. *The Quest of the Real*. Bangalore: Dharmaram Publications 1975.

KUYER, K. D. "The Problem of Ground in the Philosophy of Martin Heidegger." *The Thomist* 47 (1983) 100-117.

LANGAN, T. *The Meaning of Heidegger*. New York: Columbia University Press 1959.

LAURA, W. "Let it Be: Heidegger and future Generations." *Enviornmental Ethics* 7/4 (1985) 341-350.

LOTZ, J. B. *Martin Heidegger und Thomas von Aquin: Mensch, Zeit, Sein*. Freiburg: Herder 1975.

LÖWITH, K. "Martin Heidegger and F. Rosenzweig on Temporality and Eternity." *Philosophy and Phenomenological Research* 3 (1943) 53-77.

------------. *Heidegger: Denker in dürftiger Zeit*. Göttingen:

Vandenhoeck und Ruprecht 1965.

LYNCH, L. "Martin Heidegger: Language and Being." in *An Etienne Gilson Tribute*. ed. O'NEIL, C. J. Milwaukee: The Marquette University Press 1959.

MACANN, C., ed. *Martin Heidegger: Critical assessments, Vol.III: Language* (The Routledge critical assessments of leading Philosophers). New York: Routledge 1992.

MACQUARRIE, J. *Heidegger and Christianity*. London: SCM Press Ltd. 1994.

MAGNUS, B. "Heidegger and the truth of Being." *International Philosophical Quarterly* 4 (1964) 245-264.

------------. *Heidegger's Meta-history of Philosophy: Amor Fati, Being and Truth*. The Hague: Martinus Nijhoff 1970.

MALIK, C. H. "A Christian Reflection on Martin Heidegger." *The Thomist* 41 (1977) 1-61.

MALY, K. "From Truth to Alētheia to Opening and Rapture." *Heidegger Studies* 6 (1989) 27-42.

MARX, W. *Heidegger and Tradition*. trs. KISIEL, T. & GRENE, M. Evanston: Northwestern University Press 1971.

MAY, R. *Heidegger's Hidden Sources: East Asian Influences on his Work*. London: Routledge 1996.

MEGILL, A. *Prophets of Extremity: Nietzsche, Heidegger, Foucault, Derrida*. London: University of California Press 1985.

McCORMICK, P. J. "Interpreting the Later Heidegger." *Philosophical Studies* 19 (1970) 83-101.

MEHTA, J. L. "Heidegger and Vedanta: Reflections on a Questionable Theme" *International Philosophical Quarterly* 18 (1978) 121-149.

------------. " Heidegger and the Concept of Indian and Western Philosophy." *Philosophy East and West* 20 (1970) 303-317.

------------. *Martin Heidegger: The Way and the Vision*. Honolulu: The University of Hawaii Press 1967.

------------. "Heidegger and the comparison of Indian and Western Philosophy." *Philosophy East and West* 20 (1970) 303-317.

MITCHELL, D. W. "Commentary on Elisabeth Feist Hirsch's "Martin Heidegger and the East." *Philosophy East and West* 20 (1970)

265-269.

MOHANTY, J. N. "Consciousness and Existence: Remarks on the Relation between Husserl and Heidegger." *Man and World* 11 (1978) 324- 335.

------------. "Phenomenology and Existentialism: Encounter with Indian Philosophy." *International Philosophical Quarterly* 12 (1972) 173-184.

MORRISTON, W. "Heidegger on the World." *Man and World* 5 (1972) 452-466.

MUGERAUER, R. *Heidegger's Language and Thinking*. New Jersey: Atlantic Highlands Humanities Press 1964.

MULLIGEN, K., ed. *Language, Truth and Ontology*. Dordrecht: Kluwer 1992.

MUNSON, T. N. "Heidegger's Recent Thought on Language." *Philosophy and Phenemenological Research* 21/3 (1961) 361-372.

MURRY, M., ed. *Heidegger and Modern Philosophy: Critical Essays*. London: Yale University Press 1978.

O'CONNER, T. "Heidegger and the limits of Language." *Man and World* 14 (1981) 3-14.

OKRENT, M. B. "The Becoming of Being." *Man and World* 11 (1978) 281-298.

PARKES, G., ed. *Heidegger and Asian Thought* . Honolulu: The University of Hawaii Press 1987.

PASLICK, R. H., "The Ontological Context of Gadamer's "Fusion": Boehme, Heidegger and Non-Duality." *Man and World* 18 (1985) 405-422.

PAX, C. "Companion Thinkers: Martin Heidegger and Saint John of the Cross." *Philosophy Today* 29 (1985) 230-244.

PÖGGELER, O. "Metaphysics and Topology of Being in Heidegger." *Man and World* 8 (1975) 3-27.

------------. "Historicity" in Heidegger's Late Work."*Southwestern Journal of Philosophy* 4/3 (1973) 53-73.

------------., ed. *Heidegger: Perspektiven zur Deutung seines Werks*. Köln: Kiepenheuer & Witsch 1969.

------------. *Der Denkweg Martin Heideggers*. Pfullingen: Neske 1963.

POWELL, R. A. "The later Heidegger's Omission of the Ontic - Ontological Structure of Dasein." in *Heideger and the path of Thinking*. ed. SALLIS, J. Pittsburgh: Duquesne University Press 1970. 116-137.

------------. "Has Heidegger destroyed Metaphysics?" *Listening* 2 (1967) 52-59.

RAMPLEY, M. "Meaning and Language in early Heidegger: From Duns Scotus to Being and Time." *The Journal of British Society for Phenomenology* 25/3 (1994) 209-228.

RAUCHE, G. A. "Back to Metaphysics." *Man and World* 10 (1977) 393-409.

RICHARDSON, W. J. *Heidegger: Through Phenomenology to Thought*. The Hague: Martinus Nijhoff 1963.

------------. "Heidegger and the Problem of Thought." *Revue Philosophique de Louvain* 10 (1962) 58-78.

------------. "Martin Heidegger: In Memoriam." *Man and World* 10 (1977) 6-12.

------------. "Heidegger and God- And Professor Jonas." *Thought* 40 (1965) 13-40.

------------. "Heidegger and the Origin of Language."*International Philosophical Quarterly* 2 (1962) 404-416.

ROBERT, B. *The Question of Language in Heidegger's History of Being*. Atlantic Highlands, NJ: Humanities Press 1985.

ROBINSON, J. M. & COBB, J. B., eds. *The Later Heidegger and Theology*. New York: Harper 1963.

SALLIS, J. C. "Phenemenology and Language." *The Personalist* 48/4 (1967) 490-508.

------------. *Heidegger and the Path of Thinking*. Pitttsburgh: Duquesne University Press 1970.

------------. "Language and Reversal." *Southern Journal of Philosophy* 8 (1970) 381-397.

SCHEHAN, R. W. & MOHANTY, J. N., eds. *Thinking about Being: Aspects of Heidegger's Thought*. Norman University: Oklaham Press 1984.

SCHRADER, G. A. "Heidegger's Ontology of Human Existence." *The Review of Metaphysics* 10 (1956) 35-56.

SCHÜRMANN, R. "Heidegger and Meister Eckhart on Releasement." *Research in Phenomenology* 3 (1973) 95-119.

SHEEHAN, T., ed. *Heidegger: Man and Thinker*. Chicago: Precedent Publishing 1981.

------------. "Heidegger, An Introduction to the Phenomenology of Religion." *The Personalist* 60 (1979) 3-20.

SONTAG, F. "Heidegger, Time and God." *The Journal of Religion* 47/4 (1967) 279-294.

SPIEGELBERG, H. *The Phenomenological Movement*, Vols. I &II. Hague: Martinus Nijhoff 1960.

SPROKEL, N. *Das Nichts und die Negativität nach Martin Heidegger*. Roma: Pontificiae Universitatis Gregorianae 1972.

STACK, G. J. "The Language of Possibility and Existential Possibility."*The Modern Schoolman* 50 (1973) 159-192.

STAMBAUGH, J. "An Inquiry into Authenticity and Inauthenticity in Being and Time." *Research in Phinomenology* 10 (1979) 130-133.

STANDISH, P. *Beyond the Self. Wittgenstein, Heidegger and the limits of Language*. Aldershot Hampshire: Avebury Publishing Co. 1992.

STENLUND, S. *Language and Philosophical Problems.* London: Routledge 1990.

STENSTAD, G. & KENNETH, M. "Language as Saying: The Way of Gathering and Disclosure." *Philosophy Today* 30 (1986) 126-136.

TAYLOR, C. "Theories of Meaning." *Man and World* 13 (1980) 281-302.

VADAKETHALA, F. J. *Discovery of Being*. Bangalore: Dharmaram College 1970.

VALLICELLA, W. F. "Heidegger's Reduction of Being to Truth." *The New Scholasticism* 59 (1985) 156-176.

VERSÈNYI, L. *Heidegger, Being and Truth*. New Haven: Yale University Press 1965.

VYCINAS, V. *Earth and Gods: An Introduction to the Philosophy of Martin Heidegger*. The Hague: Martinus Nijhoff 1961.

WALHENS, DE. A. & BIEMEL, W. "Reflections on Heidegger's

Development: *Apropos* a Recent Book." *International Philosophical Quarterly* 5 (1965) 475-502.

WALSH, R. "The healing Word. Language, thinking and Being in the earlier and later Philosophy of Heidegger." *Philosophy Today* 35/ 3,4 (1991) 228-238.

WALTER, J. S. "Heidegger and Jacob Grimm on Dwelling and the Genesis of Language." *Modern Schoolman* 62 (1984) 43-51.

WELTE, B. "God in Heidegger's Thought." *Philosophy Today* 26 (1982) 85-100.

WHITE, C. J. "The Time of Being and the Metaphysics of Presence." *Man and World* 29 (1996) 147-166.

WHITE, D. A. *Heidegger and Language of Poetry*. Lincoln 1978.

------------. "On the limits of language: A critique of Heidegger and Strawson." *Independent Journal of Philosophy* 5 (1988) 135-141.

------------. "Truth and Being: A critique of Heidegger on Plato." *Man and World* 8 (1975) 118-134.

WILL, F. "Heidegger and the God of Poetry." *The Personalist* 43/2 (1962) 157-167.

WILLIAMS, J. R. *Martin Heidegger's Philosophy of Religion*. Waterloo: Wilfrid Laurier University Press 1977.

WOLI, R., ed. *Heidegger Controversy: A Critical Reader*. New York: Columbia University Press 1993.

WOOD, D., ed. *Heidegger and Language*. Warwick: University of Warwick Coventry 1981.

WOOD, R. E. "Silence, Being and the Between: Picard, Heidegger and Buber." *Man and World* 27 (1994) 121-134.

YEOW, C. L. *An Asian Looks at Martin Heidegger*. Singapore: Stamford College Press 1977.

ZIMMERMAN, M. E. *Eclipse of the Self: The Development of Heidegger's Concept of Authenticity*. London: Ohio University Press 1981.

INDEX